MUSCLECAR
COLOR • HISTORY

MERCURY
MUSCLE CARS

David Newhardt

MBI Publishing Company

Dedication

This book is dedicated to Randy Leffingwell,
who knew I could do it,
even when I didn't.

First published in 1999 by MBI Publishing Company, 729 Prospect Avenue, PO Box 1, Osceola, WI 54020-0001 USA.

© David Newhardt, 1999

MBI Publishing Company books are also available at discounts in bulk quantity for industrial or sales-promotional use. For details write to Special Sales Manager at the Motorbooks International Wholesalers & Distributors, 729 Prospect Avenue, PO Box 1, Osceola, WI 54020-0001 USA.

Library of Congress Cataloging-in-Publication Data
Newhardt, David.
     Mercury muscle cars / David Newhardt.
          p.   cm. — (Muscle car color history)
     Includes index.
     ISBN 0-7603-0549-8 (pbk. : alk. paper)
     1. Mercury automobile—History. 2. Muscle cars—History.
     1. Title. II. Series: MBI Publishing Company muscle car color history.
TL215.M425N48   1999
629.222'2—dc21   99-32415

On the front cover: The Cyclone was built on the Comet platform and made its first appearance in 1964. Eventually, the Comet platform was dropped, and the intermediate-sized Cyclone became a cousin to the Ford Torino. For 1970, the Cyclone underwent some significant styling changes. And for the first time, the 429 Cobra Jet, which pumped out 360 horsepower at 4,600 rpm, was standard on the Cyclone.

On the frontispiece: The attractive rear deck spoiler was one of many intriguing touches on the Larry Shinoda–styled 1970 Cougar Eliminator. The classic, graceful lines of this top-optioned, high-performance hardtop include a front as well as a rear spoiler; blacked-out hood scoop, front grille, and taillight panel; color-keyed sideview mirrors; side stripes; plus a special stripe treatment on the hood bulge and the rear wing spoiler.

On the title page: Heavy-metal muscle—the 1963 Marauder convertible, Mercury's version of the Ford Galaxie XL, tipped the scales at 4,043 pounds, and a bevy of big-block engines helped power the full-sized boulevard cruiser to acceptable performance figures. It had a 390-ci two-barrel base engine while four-barrel 390s, 406s, and 427s were optional. A dual-quad Holley-equipped 427 was the top option, producing 425 horsepower, which propelled the car to respectable performance figures.

On the back cover: This 1970 Eliminator wears Darth Vader black, a rare color for Mercury's Boss and Mach 1 equivalents. The Eliminators were the highest high-performance Cougars and were fitted with every hot engine Ford made: the Boss 302, 351 Cleveland, 390, and 428 Cobra Jet. These cars lit up the streets with spectacular performance, and a plush ride that distinguished them from their Ford siblings.

Edited by Paul Johnson
Designed by Todd Sauers
Printed in Hong Kong

Contents

Acknowledgments

No book is truly a solo endeavor, and without the help of a large number of individuals, this project would never have happened. I extend a heartfelt thank you to all of the people who showed me that Mercury has a valuable story to tell.

Jack Telnack was able to put the company into perspective and opened doors that I didn't even know existed. Parnelli Jones, "Dyno" Don Nicholson, Cale Yarborough, John Aiken, Darryl Behmer, Robert D. Negstad, and Curtis "Crawfish" Crider endured seemingly pointless questions to bring Mercury's history to life. C. Gayle Warnock showed me that having a great product is only half the story; promoting it is the other half.

A book about automobiles would be incomplete without photographs, and these Mercury owners were kind enough to allow me access to their vehicles. John Adams, Michael and Jennifer Baumann, John Baumann, Kenn Funk, Richard Ladd, Rob Morrow, Eugene Pokorny, Valerie Sweatte, Alex Takessian, Ray Turner, John Waugh, and Marlin "Dutch" Zeinstra put up with my crazed efforts to do justice to their cars. Kim Bray and Dan Pfeiffer of Pfeiffer Lincoln-Mercury in Grand Rapids, Michigan, went the extra mile to accommodate a tight shooting schedule. Marc Bodrie and the Cougar Club of San Diego showed me what a wonderful car Mercury produced.

Acquiring information can be the hardest part of writing a book, but these people helped more than words can say. The folks at the Ford Motor Company, Media Information Center, Theresa R. Agius and Scott C. Davis, were invaluable, as were Pat Jeffries and Buz McKim at NASCAR (National Association of Stock Car Auto Racing) Archives. John Clinard and Sandra Badgett of Ford Motor Company, Western Region Public Affairs, put me in touch with the people that I needed, even if I didn't know I needed them. Guy Preuss and Cliff Pritchard at the San Diego Automotive Museum library helped me pour through countless back issues of magazines. Mickey McGuire, visual inspiration and co-founder of Boulevard Photography, helped me understand what it was like to share a studio with a full-grown cougar. Robert Genat and Gary Jankowski provided me with unpublished, original photography, something every author dreams of.

Janis Durelle never stopped pushing when it was needed, which was often. Thank you. My editors, Paul Johnson and Zack Miller, who let me learn in my own fashion, exhibiting far more patience than I would have in their shoes. Thanks to the members of Motor Press Guild, who supported this fledging writer and welcomed me into "the business." To my sons, Branden and Ryan, thanks guys, for putting up with their dad getting "just one more." Last but by no means least, thanks to D. Randy Riggs, Matt Stone, and Larry Blankenship for giving me the chance in this wonderful arena.

—David Newhardt

Birth of the Winged Messenger
Mercury's Early High-Performance Years

Speed sells. This has been a fact of life in Detroit since the first self-propelled buggy clattered down the cobblestones in the late 1800s. Henry Ford built and raced his own creation, the famed "999," which Ford drove on the ice of Lake St. Clair January 12, 1904. He set a land-speed record with "999," covering one mile in the blistering pace of 39.4 seconds. Ford knew that performance helped sell cars, as famed race car driver Barney Oldfield drove "999" around the country, getting the Ford name in the public's mind. When the Model T was introduced in 1908, the low price and Ford's prior racing activities helped the vehicle to become a success. When it comes to the Model T, success is an understatement. That car created the Ford dynasty and changed the American automotive landscape forever. He kept his hand in racing all his life, and Fords showed up in the winner's circle with increasing frequency. Throughout Ford Motor Company's history, it has been committed to high performance and using racing as research and development for producing better cars. Use of Ford products in a pace car capacity, such as the 1935 Ford V-8 at the Indianapolis 500, raised the profile of the entire line.

With the introduction of the Mercury Division in 1939, Ford Motor Company filled what it perceived as a gap in the sales "ladder." Edsel Ford created the Mercury Division to fit directly between the Ford Deluxe and the Lincoln-Zephyr. It was intended to be a stepping stone upwards, much like General Motors hierarchy from Chevrolet to Cadillac. Fitted with a 95-horsepower flathead V-8, it resembled its Ford

The 1957 Turnpike Cruiser Convertible was powered by a 290-horsepower, 368-ci V-8. It topped out at 120 miles per hour and weighed in at 4,100 pounds. Its luxury appointments and plush suspension calibration provided a comfortable ride.

This car is the genesis of Ford's racing involvement. Henry Ford's *999* land-speed record car was driven on frozen Lake St. Clair in January 1904. It covered one mile in 39.4 seconds. *Ford Motor Company*

"cousins," but with up-market features such as semi-skirted wheel openings and an increase in body width of 6 inches. This would not be the last time that Mercurys would grow from Ford offerings.

Performance was not a priority at Mercury in its early years. Instead, establishing itself in the market and pulling buyers away from rivals was paramount. Mercury sold 69,135 vehicles in 1939, its first year of production, and by the end of the 1940 model year, 86,685 cars found customers. Most importantly, the Mercury line was accepted by the public, and the future looked good for Ford Motor Company's latest division. Then World War II intervened and halted private automobile production. Mercury resumed production on November 1, 1946, basically selling warmed-over versions of the 1942 model. The pent-up demand for automobiles meant that whatever could be built would be bought. An over-riding demand to produce an all-new Mercury didn't exist. So until the 1949 "James Dean" Mercurys were introduced, the public snapped up vehicles that advertisements described as "Graceful as a yacht!" and noted that the Mercury was "Famous among young moderns as the smart car." Young moderns indeed.

One of the earliest Mercury's used for performance or competition was the 1948 Sedan-Coupe raced by Troy Ruttman and Clay Smith. They bought it from a used car dealer in 1952 for $1,000. Besides boring the engine block and installing Edlebrock heads and an Edmunds twin-carburetor manifold, the car was basically stock. They entered it in the 1952 Panamerican Road Race, finishing fourth behind the well-financed Hemi-powered Chrysler New Yorker of Bill Sterling, and the two Ferraris that came in first and second. The full-sized Mercury could get up to 115 miles per hour, a considerable feat when you consider the bulk. This would not be the last time that the mid-range offering from the Ford Motor Company would acquit itself well in competition.

James Dean Redefines Mercury's Image

Closing the first half of the twentieth century, Mercury introduced a new vehicle. Before the 1949 models were introduced, Mercurys had been Fords with different bodywork, upscale trim, and improved interior treatment. With the release of the "James Dean" bathtub models, FoMoCo changed Mercury's direction and image—forever. With the 1955 release of *Rebel Without a Cause*, starring James Dean, this body style gained youth market appeal. Now it was much closer to the entry-level Lincoln. It was, frankly, large. Seating six comfortably, it was, sales literature noted, "longer, lower, wider." However, what was under the hood was not new. Still using the redoubtable flathead V-8, the ad writers crowed about the 1949 Mercury: "It's got plenty of get-up-and-go!" With a compression ratio of 6.8:1 helping deliver 110 horsepower from 255 ci, the leviathan could be coaxed up to 95 miles per hour, and torque was rated at 200 foot-pounds at 2,000 rpm. The public sure loved it, as did Mercury salesmen, who wrote up 301,319 sales orders, three times what had sold the prior year.

For the first time, a Mercury was the pace car at the 1950 Indianapolis 500. The yellow convertible was guided around the Brickyard by Benson Ford, Mercury general manager of the Mercury Division. Mercury also did well in the Mobil Grand Canyon Economy Run, which was sponsored by *Motor Trend* in 1950. The six-passenger Coupe tied with Studebaker at 26.6 miles per gallon. But a Mercury, prepared and driven by Bill Stroppe, was declared the winner on the basis of its ton-mpg ratio.

The checkered flag waved for Mercury in 1950 as the marque won a pair of NASCAR (National Association for Stock Car Automobile Racing) stock car races. The biggest was the 250-lap race at the 5/8-mile Oakland Speedway, where Marvin Burke took the checkered flag with an average speed of 77 miles per hour. In May 1950, *Motor Trend* tested the Sport Sedan with the 4.27:1 axle ratio and Touch-O-Matic overdrive. It took the sedan 15.98 seconds to see 60 miles per hour on the speedometer. With a top speed of 83.75 miles per hour, it wasn't going to set the world on fire, but it was comfortable, solid transportation for burgeoning interstate highway system, and after all, the top speed and performance of the car were competitive with other makes of the day.

Ad copy for the 1951 Mercury lineup was rather reserved about its "Hi-Power Compression" powerplant, encouraging prospective buyers to "Test the whisper-hustle of its great 8-cylinder, V-type engine." Though it retained the same 6.8:1 compression, Mercury engineers coaxed 112 horsepower at 3,600 rpm. Torque went up as well, to 206 foot-pounds at 2,000 rpm. One engineering feature that was new for Mercury in 1951 was the Merc-O-Matic Drive automatic transmission. Developed in conjunction with Borg-Warner, Mercury got first call on this autobox, even before Ford. The same transmission, when installed in Ford products, was cleverly called Ford-O-Matic.

Tom McCahill of *Mechanix Illustrated* drove a 1951 Sport Sedan down the quarter mile, recording a 0-60 time of 16 seconds and needing 20.8 seconds to cross the finish line at 69 miles per hour. Once again, Mercury entered the Mobilgas Economy Run. Finish-

Ford introduced the Mercury Division in 1939, and this was its sports offering for that year—the two-door Coupe, which sold for $957. Its 95-horsepower flathead V-8 provided respectable performance for the 3,000-pound car. *Ford Motor Company*

ing third provided the copywriters with fodder for their ads, noting that as a buyer "you're smart to put your money in this penny-pincher." Sales figures for that year show that 310,387 units went out the door, which made 1951 the second best sales year in its short history. Economy was the watch-word, not performance. But that would come, eventually.

Into the 1950s with Wings, Chrome, and Audacious Styling

The copywriters probably had a ball penning the 1952 advertisements. When was the last time you read about "sweeping, jet-lined grace," "Air Foil" feature lines, or a "Jet-Scoop" hood? The proud owner would gaze over the "Interceptor" instrument panel, bristling with "joy-stick" controls, confident that they were behind the wheel of "America's No. 1 Styling Car." Such hyperbole could be excused, as the all-new body shell was quite a visual departure from the prior generation. What Mercury ads called "Forerunner Styling" was in fact a celebration of American ideals. Flashier than in previous years, the chrome-laden trim and bodywork exuded optimism. Mercury shared the new body shell with Ford, unlike the 1949-1951 models, in which Mercury shared Lincoln's body shell. Some would say that the 1952 Mercury was an improvement, others, such as Ted Koopman of *Speed Age* said, "If credit lines were customary in the automobile industry, the Mercury's would include: Air scoops by Cadillac; taillights courtesy of Oldsmobile; windshield by Studebaker; and grille dentures by Buick." Still, there was no denying that a forward look had found its way into Mercury's studios. The ad agency had to emphasize the stylistic changes, because there were not many under the hood.

One of the pivotal events in Mercury racing history took place at the 1952 Panamerican Road Race. A 1948 Sedan-Coupe like this was driven by Troy Ruttman and Clay Smith to a fourth place finish. Mercury garnered a lot of attention from the performance. *Ford Motor Company*

Mercury planned to have a large, functional hood scoop accompany the release of the new overhead valve V-8 engine. Unfortunately, the engine was not ready for release. The tried and true flathead V-8 was retained, so the hood scoop air inlet was blanked off, and the flathead's air cleaner filled the space under the lowered hood. The compression ratio was bumped to 7.2:1, which helped eke out a few more horsepower. The flathead churned out 125 horsepower at 3,700 rpm, torque increased to 211 foot-pounds at 2,100 rpm, and a weight reduction of 100 pounds helped the car retain a "snappy" throttle response. Mercury ad agency covered the fact of the missing new engine with a clever spin. "There's something our engineers knew you never wanted changed. And they were able to keep it and increase horsepower (12 percent) and compression at the same time." The "Centri-flo" carburetor and the Holley 1901FF dual downdraft model helped attain the massive power increase. And just what sort of tangible performance numbers were the result of all of this technology? *Speed Age* tested a 1952 four-door Sedan, equipped with overdrive, and went from rest to sixty in 13.57 seconds, topping out at 97 miles per hour. Bill Stroppe and Clay Smith won their class in the 1952 Mobilgas Economy Run with a mileage figure of 59.7117 ton miles per gallon, or 25.4093 standard miles per gallon. More grist for the copywriters. But as far as performance went, Mercury could only get so much out of the old flathead.

The following year was the 50th anniversary of the Ford Motor Company. Therefore, every car it built that year had a commemorative steering wheel horn button. The "Unified" styling was promoted, and buyers were able to choose from a vast array of colors, fabrics, and options. Unfortunately, there were no options for the space under the hood. The three-main-bearing, 255.4-ci flathead V-8, blessed with the same 125-horsepower rating was used because the new overhead-valve engine wasn't ready to hit the streets. *Motor Trend* tested a four-door Sedan equipped with Merc-O-Matic, and the results were pretty much the same. It took 19 seconds for the sedan to get 60 miles per hour. It clipped the lights through the quarter-mile in 20 seconds, and the average top speed was 88.52 miles per hour. It wasn't much of an improvement from a performance standpoint, but sales were spectacular; 305,863 Mercurys were sold. Why mess with success?

Finally, the new engine arrived in model year 1954. Called the V-161, it had aluminum alloy pistons, "Twin Tornado" combustion chambers with 7.5:1 compression and a Holley model 2140 four-barrel carburetor. It had a 3 5/8 bore and a 3 3/32 stroke and 256 ci of displacement. Output was advertised at 161 horsepower at 4,400 rpm, while 238 foot-pounds of torque were available from 2,000 to 2,800 rpm. This woke people up. In the May 1954 issue of *Motor Trend*, the Mercury Monterey four-door Sedan with a Merc-O-Matic was wrung out. The numbers look a bit more interesting than preceding years. The 3,850-pound sedan went from 0-to-60 in 14.9 seconds, flashed down the quarter-mile in 19.4 seconds, and crossed the finish line at 69 miles per hour. Four runs for top speed averaged 97.7 miles per hour. Compared to its predecessors, this was high performance!

Also new for 1954 was an option that emulated Vista-Dome trains winding through the vast Western reaches. The Mercury Sun Valley debuted, featuring a transparent forward roof section made of Plexiglas tinted a neutral blue-green. It allowed occupants to work on their tan while enjoying the all-weather security of a closed car. All that for only $2,582. But like any Mercury of the era, lofting down America's growing Interstate Freeway system was its *raison d'être* (reason for being). Its turning circle of 41 feet, and five turns of the steering wheel lock-to-lock, give you an idea of this vehicle.

A freshened body was the news on Mercurys for model year 1955. The roof line was lowered on some of the models, and a "Full-Scope" wraparound windshield was installed. Fender treatments were varied as the division was trying to establish its own, separate visual identity that wouldn't remind people of a large Ford or a small Lincoln. Once again, the subtle copywriters working for Mercury had their fingers on the pulse: "Exclusive styling—You don't have to look twice to tell it's a Mercury."

The Montclair series was introduced at the Chicago Automobile show in January 1955. The Sun Valley hardtop was incorporated into the Montclair mix, as was the only convertible in the entire Mercury lineup. The wheelbase stayed at 118 inches, and the standard tires, 7.10x15, were a tubeless design, for the first time. But what turned those tires was a pair of "Super-Torque" V-8s, thanks to Mercury engine guru Vic Raviolo. The Y-block engine was revamped and incorporated a score of improvements. A new cylinder-head design, featuring an "open-wedge" profile, revised porting and a high-lift camshaft were just a few of the enhance-

With all new styling, the 1949 Mercury would find favor with James Dean and the public. Dean drove the car in the movie *Rebel Without a Cause*. The success of this all-time classic movie translated into higher visibility and greater success for Mercury. *Ford Motor Company*

ments. The five-main-bearing engines had redesigned valve covers, modified timing chain, and a newly designed air cleaner, all in an effort to reduce noise. The valvetrain featured adjustable solid lifters with adjusting screws, and the related components—valve tappets, pushrods, and the camshaft—all benefited from improved metallurgy. Heads were fitted with new, 18 mm "Turbo-Charge" Champion spark plugs. The conical-seat design reduced fouling due to a wider space between the insulator and the outside shell.

Both "revised" engines displaced 292 ci and had a bore of 3 3/4 inches and a stroke of 3 19/64 inches. Holley model 4000 four-barrel carburetors equipped with automatic choke fed the powerplants fuel. The first engine was fitted to all Montclairs, Montereys, and Customs and equipped with the manual transmission. It had a 7.6:1 compression ratio, and cranked out 188 horsepower at 4,400 rpm. Late in the 1955 model year, horsepower increased to 193 horsepower, and torque was measured at 274 foot-pounds at 2,500 rpm. But if a buyer ordered a Montclair with the Merc-O-Matic automatic transmission, the V-8's compression ratio was raised to 8.5:1. This in turn helped the engine to produce 192 horsepower at 4,400 rpm and 286 foot-pounds of torque at 2,500 rpm. Horsepower output climbed once again by late 1955, rising slightly to 198.

The 1954-55 Sun Valley was a two-year offering that brought wide-open vistas to a dazzled public. It was not a big seller, but it helped raise the profile of the "middle division" of the Ford Motor Company. *Ford Motor Company*

The Merc-O-Matic was refined to improve shifting smoothness. The engine produced so much torque that the transmission was set up to start out in second gear when the gear selector was put into drive. However, if the driver got on the gas hard, it would kick down into first gear. When installed in a Mercury, the tranny was called the MX transmission, while the Ford version was known as the FX. As an interesting aside, 1955 Thunderbirds, and "312" V-8–equipped 1956 T-Birds used the Mercury transmission. So just what kind of performance improvement did we see? *Motor Life* ran a 1955 Montclair down the quarter-mile at 19.5 seconds and achieved a 0-60 time of 11.7 seconds. A four-run average speed of 103.1 miles per hour was timed, with one of the runs clocking in at 107.9 miles per hour. *Motor Trend* had their way with a 188-horsepower, Merc-O-Matic–equipped Custom four-door Sedan, posting results that backed up what similar publications had found. With a 0-60 time of 11.4 seconds and running the length of a quarter-mile drag strip in 18 seconds at 78 miles per hour at the finish, it was quicker than the Montclair. *MT* made a four-pass top speed test, averaging 105.3 miles per hour, with the fastest one-way speed of 107.9 miles per hour.

The public obviously liked what they saw and test drove. Mercury set another all-time sales record: 329,808 units in model year 1955. They hoped to repeat it with the 1956 lineup, which was carried over intact with the exception of the Sun Valley hardtop. Three new models were introduced, including a Custom Station Wagon. A low-end version, called the Medalist, was inserted into the Custom line. A Monterey version of the Sport Sedan was introduced as well. The economy at the time was in a mild recession, and Mercury was able to gain substantial market share, while its rivals at General Motors lost a significant portion of the market. The "entry-level" Medalist helped a lot of people, almost 40,000, into a Mercury. And to get those cars down the road were new Mercury's tag line "Safety-Surge" V-8s. These engines could "deliver that extra margin of power where and when you need it—for split-second pickup, safer passing, easier hill climbing." A foursome of engines were offered, three for regular street use, the fourth if your regular street was a stock car racing track.

All four engines displaced 312 ci and had a new combustion chamber design. Larger piston rings, precision-molded crankshafts, and a change to a 12-volt electrical system were a couple of the engineering changes that Mercury incorporated. Most of the engines destined for street use used a Holley Model 4000 four-barrel carburetor with an automatic choke and idling control. If the vehicle was equipped with an automatic transmission or power brakes, the carb was equipped with a dashpot. While some cars came equipped with a Carter WCFB number 2361S four-barrel carburetor, most were fitted with the Holley. If the Mercury carried an overdrive transmission or the three-speed manual, the mill under the hood developed 210 horsepower at 4,600 rpm. With a compression ratio of 8.0:1, it produced 312 foot-pounds of torque at 2,600 rpm. The second engine had 8.4:1 compression, 220 horsepower at 4,600 rpm, and 320 foot-pounds of torque and was installed on vehicles that used the Merc-O-Matic. An optional engine for use with the Merc-O-Matic transmission produced 225 horsepower at 4,600 rpm and 324 foot-pounds of torque at 2,600 rpm and had 9.0:1 compression ratio. The M260 engine kit that dealers installed was another option available to Mercury faithful. New cylinder heads, camshaft, and an intake manifold topped with a pair of Holley four-barrels coaxed 235 horsepower out of the Y-block. Not widely advertised, this was intended for driving in left-hand circles.

Mercury was starting to establish its presence in NASCAR's Grand National series. Tim Flock got the ball rolling for Mercury when he took Bill Stroppe's M335 to the checkered flag at Road America in Elkhart Lake, Wisconsin, in 1956. His M355 was equipped with a pair of four-barrel carburetors on top of a 368-ci Y- block, and two shock absorbers at each

For 1957, the Big M had beautifully sculptured projectiles for the "look of tomorrow." The production Turnpike Cruiser was 211.1 inches long and came an abundance of gadgets and distinct 1950s styling. It was powered by a 290-horsepower 268-ci V-8. *Ford Motor Company*

corner provided the dampening. The 335-horsepower Mercury was able to sit on the pole of the 1957 Southern 500 with a speed of 117.416 miles per hour.

Auto Age magazine tested the 225-horsepower Merc-O-Matic—equipped Mercury Monterey in its June 1956 issue. The numbers the car posted in the evaluation showed that the efforts of the engineers had not been in vain. The full-sized car ran 0-60 in 11.1 seconds and topped out at 105 miles per hour. Though the brakes tended to fade quickly, the Monterey displayed good handling manners. In spite of the declining economy, enough buyers thought that the Mercury line fulfilled their needs, and 246,629 automobiles rolled out of the showrooms.

The Full-Sized and Fully Optioned Turnpike Cruiser

Don DeLaRossa and his design staff at Mercury had built a show car for the 1956 auto show circuit. Called the XM Turnpike Cruiser, it was spectacular! It had a three-piece wraparound rear window and transparent roof sections. In its October 1956 issue, *Motor Trend* said that the new Mercurys would be unlike any other, and how right they were. Mercury wanted, and got, its own distinctive look that was different from Ford and Lincoln products. With the Turnpike Cruiser production models, called Hardtop Phaetons, and with the mid-year introduction of a convertible, Mercury put "horizontal plane" styling on the top-of-the-

line Interstates. With the general shape, and acreage, of an aircraft carrier, the horizontal plane styling took long, low, and wide to extremes. The wheelbase was stretched to 122 inches, and it had with an overall body length of 211.1 inches, which worked out to 17.59 feet—good luck trying to fit a Turnpike Cruiser in one of today's garages! At 79.1 inches across, Mercury gained almost 3 inches in width. This was a source of pride for Mercury, and proof that buyers were getting good value for their dollars, or at least a heck of a lot of metal.

A convertible version appeared later in the model year and paced the field for the 1957 Indianapolis 500 race. The Pace Car, prepared by Bill Stroppe, was equipped with the "Dream-Car Spare Tire Carrier," a.k.a., a continental tire kit. This tasteful option was available on all Mercurys that year. The styling was definitely pushing the envelope. The 1957 brochure informed the curious that "beautifully sculptured projectiles give the 'Big M' for '57 a definite 'look of tomorrow.' " Rockets and missiles had a strong influence on the "dream car design." A slanting retractable rear window, jutting vent ducts in the roof at the top of the hardtop's A-pillars, and a "Seat-O-Matic" memory seat with 59 positions were a few of the "space-age" enhancements that the proud owner could enjoy.

Under the hood of these rolling stylistic statements were a pair of engines that moved Mercury in

Auto show viewers in 1956 could see the future now with the XM Turnpike Cruiser. It had skylight dual windshield corners, pushbutton Merc-O-Matic transmission, a memory seat with 49 positions, and a reverse slanted retractable rear window. *Ford Motor Company*

a commanding fashion. The "Safety-Surge" 312-ci V-8 produced 255 horsepower and was the standard engine on Montclairs and Montereys. When the Turnpike Cruiser was purchased, it was fitted to a new 368-ci V-8 that had 9.75:1 compression ratio, 290 horsepower at 4,600 rpm, and 405 foot-pounds of torque at 2,800 rpm. But with a curb weight of 4,280 pounds, depending on options, every pound-foot was used to propel the Cruiser. Tom McCahill wrote in *Mechanix Illustrated* that "The 1957 Mercurys are the most different cars of the year." Regarding the Turnpike Cruiser that he tested, he noted that it had "all the roadability of a rubber-soled gazelle and could handle drifts and slides with the sureness of a competition sports car." He came up with a 0-60 time of 10.5 seconds and got the speedometer to give up 120 miles per hour. While the 368-ci engine was the standard on the Turnpike Cruiser, the powerplant could be ordered for any other Mercury. This engine incorporated a couple of engineering firsts, including a paper air cleaner and the use of hydraulic valve lifters. Despite the new look and new engines,

sales fell to 286,163 cars for 1957. The country was gripped by a severe recession. The last thing people wanted to buy was an enlarged, thirsty new car, but Mercury was confident that 1958 would see more of its metal on the road.

The big news for model year 1958 was the introduction of MEL (Mercury-Edsel-Lincoln) engines. Sister company Ford was using performance-oriented engines. The new Marauder 383 and 430-ci V-8s were slipped under the hood of the big "M." Both mills enjoyed 10.5:1 compression and required premium fuel, and lots of it. The 430-ci engine was, in fact, a 1958 Lincoln engine, the largest engine installed in any American car that year. It churned out 360 horsepower at 4,600 rpm, and it was the standard engine in the Park Lane. Later in the year, it was offered in the Turnpike Cruiser, and it would be used in Mercurys until 1960 as well as Lincolns until 1965. An excellent engine, it would purr along at freeway speeds, underworked. This engine and other engines in the family were unusual in that the top of the block was cut at a 10-degree angle relative to the top of the piston.

These over-engineered mills were designed to propel massive Lincolns. When slotted into lighter, nimbler Mercurys, these cars flat-out flew.

Tom McCahill with *Mechanix Illustrated* drove a Park Lane with a 430 engine from 0-60 in 10 seconds. The 383-ci engine was a destroked version of the 430, putting out 312 horsepower in the Monterey and 330 horsepower in the Montclair. Holley supplied the carburetors for all engines, while dual exhaust was standard if the buyer opted for the 430 engine. A "special order option" could transform a Marauder into a "Super Marauder." This 400-horsepower option was developed by Bill Stroppe, and consisted of three two-barrel carburetors. This rare engine (supposedly 100 were produced) proved to be the most powerful powerplant offered in any American car in 1958.

Sales dropped in 1957 and dived to a paltry 153,271 in 1958. The recession was taking its toll on all of the manufacturers. The only American make that posted an increase in sales was Rambler. That should have been a wake-up call, but the 1959 Mercury took lavishness to new highs. All of the styling cues from 1958 were retained, or rather magnified. The scallop at the beltline was extended forward to the base of the A-pillar. Domestic carmakers, including Mercury, seemed to have a common design rule: when in doubt, slather on more chrome. At the time when the public was starting to embrace smaller, more fuel-efficient cars and the recession was still casting a long shadow across the land, it's ironic that some of the biggest vehicles Detroit ever produced rolled onto showroom floors. However, not as many as Detroit would have liked rolled off of the showrooms. When 1959 was in the history books, Mercury had sold only 149,987 units. This was slightly worse than the prior year, but not as steep a decline as 1957 to 1958. Part of this sales malaise was due to the lineup that Mercury was offering.

The Turnpike Cruiser was gone, and Mercury changed the name of its hardtops from Phaeton to Cruiser. "Clean-Dynamic Styling" was still evident, but the rolling projectiles found that more time was required to get up to speed, as the engines had been "massaged" for improved fuel economy. In the Monterey model, an economy version with 8.75:1 compression was standard. Fitted with a two-barrel carburetor, the 312-ci engine produced 210 horsepower. The optional Marauder 383-ci V-8 with the two-barrel carb had a reduced 10.0:1 compression ratio and churned out 280 horsepower. The same size engine, but with a four-barrel carburetor, produced 322 horses. Last on the list of large V-8s was the 430-ci engine installed in the successor to the Turnpike Cruiser, the Park Lane.

This top-of-the-line Mercury's engine also suffered a decrease in its compression ratio, down to 10.0:1. With a single four-barrel carburetor, its output was 345 horsepower. The Stroppe-designed multiple carb setup was not offered; it wouldn't show that Mercury was serious about fuel economy.

Performance numbers were on a par with 1958 as *Mechanix Illustrated*'s Tom McCahill found out when he tested a four-barreled 383-equipped Colony Park Station Wagon on the Daytona race track. Although a station wagon was tested, there is strong correlation between the wagon's performance and the performance of a full-size Mercury sedan. "The Mercury handles beautifully through the corners. Even while crisscrossing slightly up and down the 31-degree bank at the Daytona track it kept grooving like a golf ball in a mail chute. The brakes are excellent, better than those on most of its competitors. I brought the car to a complete stop from 100 miles per hour in 387 feet. Top speed is definitely between 110 and 115. I averaged 111.6 miles per hour for ten miles on the Daytona Speedway. Acceleration with the single-range transmission showed 11.9 seconds from 0 to 60. The only time when discomfort is experienced is when you want to pass with a limited margin of clearance. Then you become aware, like a stood-up bride after an hour or two, that this Mercury wagon is sadly lacking in the pick-up department."

Jim Whipple of *Car Life* piloted a 430-equipped Park Lane around the Dearborn, Michigan, test track and wrote that he saw 0-60 miles per hour in just eight seconds, and that the big Merc would "leap from 50 to 70 miles per hour in a passing maneuver in just five seconds," not bad for a vehicle that tipped the scales at 4,386 pounds. Whipple continued, "I took to the bigger 'M' just like Kentucky bourbon takes to mint and ice. It's a car that a great deal of brain power has been put into."

With a palette of full-sized cars and fuel economy demands looming larger on the horizon, Mercury seemed destined to fade into obscurity. Real heart-stopping power had been available for only a couple of years, but Mercury's insistence on staying with the big cars was ill-timed for an increasingly frugal populace. Mercury, as well as virtually every other vehicle manufacturer, had been caught playing one-upmanship with the competition. It took Mercury time to respond to the changing marketplace. Massive egos were clashing, and the fallout was a score of spectacular behemoths. Correcting the damage that had been done would take time, and the introduction of the proper products for the times. But there was light on the horizon, and it looked like a Comet.

High-Performance Sizzle in a Small Package
The Comet, S-22, Caliente, and Cyclone

2

From acorns grow mighty oaks, or so goes the old saying—Mercury's first step toward a legitimate high-performance presence started with the Comet. This is not to say that when Mercury introduced the Comet for the 1960 model year, they had visions of the new re-badged Ford Falcon setting the world on fire. Mercury had been witnessing the steady erosion of market share for the last few years due to a number of factors. The economy in the late 1950s was, to put it delicately, in the throes of serious recession. The automotive buying public could not afford to purchase and operate the land-yachts of the flamboyant '50s. Smaller, economical, and more responsive cars were being snapped up in increasing numbers. The specter of imports actually pulling away domestic sales was evident. Volkswagen sold more than 55,000 Beetles in 1958, and the mood of the public was to buy small. So when the Comet debuted along with the rest of the new-for-1960 Mercury line-up, it was getting in on the ground floor of the "compact" market segment. Mercury wanted market share back, period. Ford had released its Falcon a couple of months prior to Mercury's version, and the feedback was good. However, the Mercury name didn't appear anywhere on the Comet. It was just the Comet, sold through Mercury.

But Mercury being Mercury, it would never forsake the upscale luxury that had positioned the division between Ford and Lincoln since its inception. A bit of sparkle in the compact market sector could only work to Mercury's advantage. With a wheelbase of 114 inches for the sedan and 109 inches for the station wagon, it was a dramatic reversal of the growth tendency that had happened in earlier years. The Comet sedan was 14 inches longer than its Falcon cousin, and the extra length was added to the

Mercury ventured into the sports compact market with its Comet, which was a spinoff of the Ford Falcon Futura. Though fitted later with Mustang wheels, this particular Comet, the 1963 S-22, was tastefully trimmed in chrome, had bucket seats, and a well-appointed interior.

Triple taillights made identifying an S-22 easy, as the base model Comet had two lights per side. This was a visual tie to the top range Mercurys, which had three taillights per side.

trunk. Its canted fins and tail-lights, along with the downswept trunk lid made a strong styling statement. They were not intended to burn up the road, just fulfill the public's need for comfortable, economical transportation. And with the little acorn under the hood, the world was safe again.

The Comet, an Econo Car for the 1960s

Right out of the gate, Mercury's target buyers were obvious. The inline, overhead-valve six-cylinder engine displaced 144.3 ci. With a compression ratio of 8.7:1, a single-barrel carburetor, and a horsepower rating of 90, it was hardly a threat to bigger Mercurys on the open road, but unlike its larger brethren it could deliver up to 28 miles per gallon. The three-speed manual transmission, which the vast majority of buyers purchased, helped achieve gas mileage in the high-20s. Of course, if 90 horsepower and 138 foot-pounds of torque was too much to handle, and the driver did not want to shift, a two-speed "Merc-O-Matic" automatic transmission was optional. Mercury pointed out that the engine was 32 percent lighter than the competition's six-cylinder mills. Comet had a longer wheelbase than other compacts, and it still had the famed Mercury "Big Car Ride." Jim Whipple of *Car Life* wrote about the new Comet in the April 1960 issue. He said, "This engine is smooth and well-balanced, and in the Falcon, more economical than any other compact—with the possible exception of the overdrive Lark and Rambler. However, in the Comet, the 140-ci six-cylinder engine with the same 8.7 to 1 compression ratio just doesn't have the moxie to haul the mail." But enough moved out of the showroom to

allow Mercury to post a slight sales increase in model year 1960, up to 155,000, including 116,331 Comets.

The following year the Comet was virtually unchanged, except that air conditioning was available for the first time. The Mercury nameplate didn't adorn the vehicle, but only Mercury dealers were selling the vehicle. At the New York Automobile Show, midway through the model year, the Comet S-22 debuted. Essentially, the car was a trim enhancement, which was aimed at the increasing popularity of sporty compacts. Most of the S-22 specific items were interior parts, such as the front bucket seats and the center console. Chrome wheel covers and 50 pounds of additional insulation as well as white sidewall tires were installed. If anything, the extra weight slowed the car down. But low and behold, an engine option appeared, blessed with an increase in horsepower.

By lengthening the stroke of the 144 engine by using a "long throw" crankshaft, the power-plant grew to 170 ci. A larger single-barrel carburetor, intake valves, and intake manifold helped produce 101 horsepower. The standard three-speed manual transmission was strengthened to withstand the additional power, while the two-speed automatic was offered again as an option.

Performance was not the Comet's strong suit. *Motor Life* magazine put a 1961 Comet S-22 with the 170-ci 101 horsepower engine, bolted to a Merc-O-Matic, through its paces. Their verdict: "it bogs down in performance." The car posted a 0-60 miles per hour time of 22 seconds and took 23.6 seconds to run the quarter-mile at a terminal speed of 62 miles per hour. Nevertheless, the car was a sales success, and 197,263 found customers. Economy was not ignored though, as the big-engined Comet went 16 miles around town on a gallon of regular gasoline, while pulling 20-plus miles per gallon while crossing continents.

Mercury finally acknowledged "ownership" of the Comet line by fitting them with the division name when the 1962 Comets were introduced on September 21, 1961. The biggest visual change to the line was the restyled rear end. Now, from the rear, the Comet looked like a scaled-down Mercury. The standard Comet had four taillights set into a bright, recessed, full-width trim piece. The top-of-the-line two-door sedan S-22 had six taillights and standard backup lights located in the center of the inboard two lights. Mechanically, nothing changed in 1962. The Comets had the same engines and the same 114-inch wheelbase with an overall length of 194.8 inches. With a turning circle of 39.9 feet and 4.64 turns lock-to-lock, a Comet driver kept their arms busy on a twisty road. The Comet line's compact size and 2,711-pound weight allowed it to attain the level of fuel economy. Another reason for the good mileage figures was that the vehicle was grossly underpowered and had a hard time getting out of the way of its shadow.

In May 1962, *Motor Trend* tested a Comet S-22, equipped with the "big six" engine, Merc-O-Matic two-speed transmission, air conditioning, and the optional 4.09:1 ratio rear axle. With a 0-60 time of 22.2 seconds, the reviewers were not really going out on a limb when they said that the vehicle was grossly underpowered. The "senior compact" was putting money in Mercury's bank account, to the tune of 165,305 Comets sold in model year 1962. While the number of Comets dropped from earlier years, they had the largest share of total Mercury sales. However, the economy was warming up, and buyers were starting to ask for more power under the hood.

When the 1963 Comets were released on October 4, 1962, they had the regular Detroit exterior freshening. A new S-22 Comet Convertible ("livewire

The chrome excesses of a couple of years prior had diminished with the 1963 Comet S-22. Detroit was stepping away from the "Baroque-Modern" look and embracing a more tasteful styling approach.

with the lush lines") was added to the lineup, but the rest of Mercury's efforts seemed limited to changing small bits of brightwork and installing newly styled hubcaps. Imagine the surprise and excitement generated at the midyear introduction of the new V-8 powered Comet—the 1963 Sportster. Mercury's version of the Ford Falcon Sprint was for sale. The Ford 260-ci engine slipped into the Comet chassis, and body and chassis strengthening components handled the additional power. When the V-8 came on the scene, the 144-inch six-cylinder engine was dropped, making the 170-ci "big-six" engine the standard engine. This engine came from the Fairlane/Meteor 221. With 8.7:1 compression and a two-barrel carburetor, it kicked out 164 horsepower at 4,000 rpm. The dramatic increase in torque (compared to the six-cylinder engine) was an impressive 258 foot-pounds at 2,200 rpm. As expected, performance was the beneficiary of this power transfusion. *Car Life* put a Sportster with the two-speed Merc-O-Matic through its paces and came up with a 0-60 miles per hour time of 14.5 seconds. It ran down the quarter mile in 19.3 seconds, tripping the finish line lights at 72.6 miles per hour. They also held the pedal to the floor for a top speed of 103 miles per hour. *Motor Trend* got their hands on one, too, equipped with a four-speed manual transmission. Their results were even more encouraging for fans of performance. With a 0-60 time of 11.5 seconds and a drag strip showing of 19 seconds at 75 miles per hour, Mercury showed that it was still able to produce a vehicle that was more than merely a grocery-getter. As the economy started improving, buyers were showing an interest in performance again. Mercury sold 134,623 Comets for 1963, and did not waste any time slipping yet more rotational energy under the hood.

Cyclone, a Genuine Mercury Musclecar

Racing was helping to sell the full-sized Mercurys, but the Meteor was dropped from the Mercury lineup for 1964. The Comet line had the responsibility to

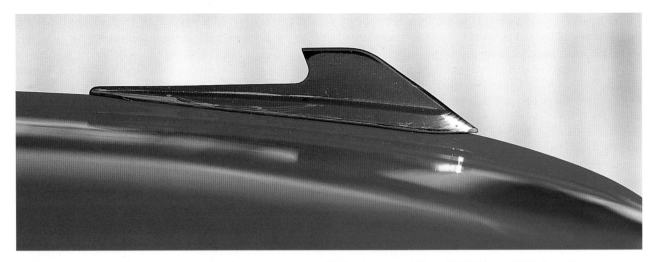

Rocket-inspired fender trim steadily disappeared as the 1960s progressed. This 1963 Comet S-22 front fender piece looked like a dorsal fin, but this kind of ornamentation was expected in that era.

increase sales to compensate. For the first time, the Comet had truly inspiring performance. With the new 9.0:1 compression 289-ci engine putting out 210 horsepower at 4,400 rpm and 300 foot-pounds of torque at 2,800 rpm, the image of the Comet changed for the better. The 90-degree V-8 thin-wall engine proved to be legendary over its lifetime, and with good reason. It responded to hop-up tricks with aplomb and proved its reliability many times. Its compact design allowed it to be used in virtually any engine compartment. Witness Carroll Shelby installing

one in an AC Ace English sports car, giving birth to the famed Cobra. With the removal of the S-22 model for 1964, the November 1963 issue of *Science & Mechanics* called the new Caliente "the hot one." The top-of-the-line Comet was forecasted to be "one of the quickest of the compacts in the stop-light grand prix's," and they got a 0-60 time of 10.5 seconds "without tricks or brutalizing the transmission."

Midway through the 1964 model year, January 17, 1964, a new name was on people's lips—Cyclone. The Caliente was no longer the top model in the

The Comet Cyclone was essentially an involved and an improved version of the S-22. From the flag on the front fender to the chrome wheels, the 1964 Comet Cyclone, with its standard 289- ci engine produced a 0-60 time of 10.2 seconds with the 4-speed manual transmission. *Ford Motor Company*

Comet lineup, and the Cyclone provided Mercury with a real high-performance car. The Cyclone would start out as a two-door hardtop only, and before it disappeared in the early 1970s, it would be one of the most successful Mercury NASCAR racers ever. "Under the hood, a whiplash of surging power" read the sales literature. The "Super Cyclone 289 V-8" was the standard Cyclone engine, and it put out 210 horsepower and 300 foot-pounds of torque. Each engine came with a chrome dress-up kit, as well as less chrome on the exterior compared to other Comets. The interior was blessed with "the masculine feel of black vinyl in the instrument panel." Quite a bit of attention was given to what was attached to the rear of the engine, the transmission. The Multi-Drive Merc-O-Matic three-speed automatic transmission was offered for the first time. The Comet actually carried the Ford Cruise-O-Matic, thus warranting the name change. A three- or four-speed manual transmission was also available for those who liked total control. An interesting note is that the 289-inch engine only weighed six pounds more than its 260-inch cousin.

The folks at *Car Life* (April 1964) tested the new Cyclone equipped with the automatic transmission, and they set a respectable 0-60 time of 11.8 seconds. Moving through the gears manually with the four-speed shaved off quite a bit of time, jumping from 0 to 60 miles per hour in 10.2. In the quarter-mile the difference was not so apparent. The autobox-equipped vehicle covered the 1/4 mile in 16.5 seconds at 73.8 miles per hour, and with the manual four-speed, the quarter-mile flew by in 16.4 seconds at 77 miles per hour. Pedal to the metal resulted in a top speed of 109 miles per hour. All of this good

Offered only as a two-door hardtop, the 1964 Cyclone's optional 289-ci, 271-horsepower engine provided plenty of power for the 2,688-pound car.

press helped 189,936 Comets find buyers in model year 1964.

For the buyer that "needed" yet more motive power, the Mercury dealers offered Shelby American Hi Performance parts, kits, and accessories. On a special-order basis was the Super Cyclone High Performance Option, which delivered a 289-ci engine pumping out 271 horsepower. From its larger four-barrel and intake manifold to the mechanical lifters and special connecting rods, the midyear option was quick to improve the Cyclone's profile. With "Street Shim Head Gaskets" selling for $5.45, to "Dual Quad High Riser" induction packages for $249.50, Cyclone owners could easily improve their vehicles' power-to-weight ratio. When these goodies first came out, the three-speed or optional Warner T-10 four-speed manual transmissions were the only ones available. But by July 1964, the three-speed Merc-O-Matic C-4 automatic transmission could be bolted to the rear of the engine block.

Creating a High-Performance Heritage

Competing and winning was the new goal for the division. To highlight the newest addition to the Comet line, the Caliente, a number of high-profile events were undertaken. The first took place at the Daytona International Speedway in Florida, starting on September 21, 1963. Andy Hotton, who had prepared the Ford Fairlane "Thunderbolt" drag cars, prepped five new Comet Calientes for an attempt on a new Class C endurance run record, which would entail driving at least 50,000 miles with an average speed of at least 100 miles per hour. Hotton built the Comets to NASCAR specifications, meaning full roll cages, heavy-duty shocks and wheel rims, and special camshafts. With 10.5:1 compression, mechanical valve lifters and oversized oil pans, they hit the track running, keeping a fast pace for six weeks. At times the Comets would see speeds of 112 miles per hour, but on October 30, 1963, they finally pulled off of the banked track. They had broken over 100 world records, while setting a pace in excess of 108 miles per hour for the full 100,000 miles.

Mercury wasted no time in touting what its durable compact had accomplished. Lincoln-Mercury Vice President and General Manager Ben D. Mills was interviewed in the September 1964 issue of Car *Life* about Comet sales following the record run. "The sales success of the 1964 Comet following the Comet Durability Run at the Daytona International Speedway last fall presents graphic proof that performance does help sell cars." He went on with hard proof. "In the third 10-day period of October, sales were up 28

he fitted them with 289-ci "Super Cyclone" engines, cranking out 271 horsepower. Other performance modifications included a 10.5:1 compression ratio, four-barrel carburetor, special camshaft, 4.57:1 rear-axle ratio, skid plates, roll bars, and metallic brake linings. The rally covered 3,188 miles and 94 vehicles started the event. Driver Don Bailey recounted what it was like driving the race.

"In road events it's not uncommon for a car to blow an engine or tear up a gearbox, but in the Safari many entrants simply collapsed into hopeless junk. In Northern Kenya, a dozen were stuck in mud so deep you couldn't open the doors. We had to push them off the road for the Comets to pass." When it was over, only 21 vehicles finished and two of the five Comets that had started were among them. One of the Mercury drivers, Ray Brock, said that the reason that the rest of the Comets had not finished was the failure of the standard American shock absorbers. But Mercury got plenty of advertising mileage from both of these events, and had more up its sleeve.

The Comet's New Clothes

It was easy to tell a 1965 Comet coming down the road. The headlights were arranged vertically, which gave the car a new distinctive look. Ford Motor Company had incorporated this look through most of its lineup, and as the Comet was the Mercury version of the Ford Falcon, it was a cost-saving measure to use the same basic substructure with new body panels. The rear was modified as well, with more of a horizontal grille look. Still riding on a 114-inch wheelbase and with an overall length of 195.3 inches, the Comets weren't genuine compacts; they were intermediates, but Mercury didn't acknowledge the fact. Under the skin, changes had been wrought. The 260-ci V-8 and 170-ci six cylinder were no longer offered. The 200-ci six was the base engine, and the "Cyclone V-8" was an option in any non-Cyclone Comet. Ordering a Cyclone resulted in all of the luxury touches of the Caliente, plus the front bucket seats with center console, tachometer, and brightwork that identified the vehicle as a Cyclone. An optional fiberglass hood was available, complete with dual air intakes. Under the hood was the 195-horsepower

A trio of 1965 Comets went from the tip of South America to Fairbanks, Alaska, to prove the durability of the vehicle and get exposure for the new, stacked headlight treatment for 1965. Few vehicle problems were encountered in the 40-day journey. Mercury got plenty of mileage out of this demonstration. *Ford Motor Company*

percent. After that, sales continued to gain momentum, reaching a high of 187 percent ahead in the final 100-day period in January." Refreshing words from a "suit."

Following this success, Mercury had Bill Stroppe prepare a handful of new Comets to participate in the "East African Safari Rally," which was held between March 26 and March 30, 1964. Stroppe went through the cars with his normal thorough fastidiousness, and

The base 1964 Comet 202 series had a minimum of chrome, and its clean lines were several years ahead of its contemporaries. While the six-cylinder engine was no screamer, it was reliable and durable.

289-ci "Cyclone V-8." As an option on all Comets, $45.20 extra with Cyclones and $153.20 extra with the rest of the Comet lineup, was the "Super Cyclone 289." With its 10.0:1 compression and four-barrel carburetor, it was good for 220 horsepower. *Motor Trend*'s May 1965 issue put a Cyclone with this hot engine through its paces and came up with a 0-60

The triple taillight treatment gave the base 1964 Comet some cachet from the high-line models. Styling was moving away from excesses of the 1950s. A much cleaner, simpler look with less chrome was being implemented. The rear grille copied the front's texture, and tail fins were slowly fading away.

time of 8.8 seconds, while finding the end of the drag strip only took 17.1 seconds, finishing at 82 miles per hour. The magazine complained about the rear axle hop and wheelspin. Despite those drawbacks, they felt that it was a step in the right direction.

Mercury was not finished with its high-visibility "runs" for its biggest seller. On September 12, 1964, a trio of 289-equipped Comets started their way north from Ushuaia, Argentina, on another factory-backed durability run, this time called the "Journey from the bottom to the top of the world." From the tip of South America to Fairbanks, Alaska, the three snow-tired vehicles wound their way north 16,247 miles in 40 days. Except for four flat tires, the Comets had no mechanical problems, handing Mercury another opportunity to extol the virtues of Comet ownership. The Lincoln-Mercury Division published a 16-page booklet that depicted the Comet in a number of extreme situations. The booklet proved the cars were durable. And they did sell. While Comet sales dropped to 154,312, the rest of the Mercury Division posted impressive increases with another all-time record of 346,751 cars for 1965. In the mid-1960s, Mercury was poised to make a far better high-performance statement than it had going into the beginning of the decade.

Mercury Builds Its Musclecar Foundation
Marauder, S-55, and X-100

3

In 1961, America was full of hope. John F. Kennedy was in the White House, the economy was in an upswing, and Mercury was ready to make a full-fledged assault on the high-performance market. Decisions had been made that would bring the Mercury line more closely into the Ford camp, away from a stand-alone division, at least stylistically. The "land-yachts" of the late 1950s, resplendent in acres of sheet metal and chrome, would not play to the people in Peoria anymore. The 1960 models, especially the Monterey, looked like something that would actually be found in a garage, not at a spaceport in a science-fiction movie waiting to take off. Okay, so the roof section looked like it was a carryover from 1959, but the body below the glass was much more subdued. Mercury let what was under the hood do the talking. That meant performance, and Mercury was about to embark on a race for performance that had never been seen before and will probably never be seen again.

Marauder, Brawn with Bulk

When the 1960 Mercury Monterey was unveiled, it was quite a departure from the preceding year, as far as the body went. The engines offered had changed also, and alas, not for the better. The strongest engine available in the Monterey in 1959 was the 430-ci V-8, putting out 345 horsepower and 460 foot-pounds of torque with a four-barrel carburetor. In 1960, the 430-ci engine was still around, but its power rating had slipped to 310 horsepower, courtesy of 10.0:1 compression and a "gas-saving" two-barrel carburetor. *Motor Life* tested a Montclair, which outweighed a Monterey by some 200 pounds, depending on options. With a 2.71:1 axle ratio, the Montclair dashed from 0 to 60 miles per hour in 11.5 seconds, this with a 2.71:1 axle ratio.

Though the 1963 Marauder convertible tipped the scales at 4,043 pounds, the well-proportioned design avoided a massive look. But the driver going into a turn too fast quickly learned that physics would not be denied.

27

The 1963 Mercury Marauder was a midyear release, riding on a 120-inch wheelbase. The standard engine was the Marauder Super 390 V-8, which pumped out 300 horsepower.

But the outlook for performance fans was upbeat with the introduction of the 1961 models.

Once again, a Ford body provided a starting point for Mercury. Big "M" went to work on the body by shortening it 4 inches and narrowing by 1 inch. Trim levels remained at Mercury upscale standards, and the engine compartment was not ignored. While this was the first time a six-cylinder engine was offered in a full-sized Mercury, a new name, Marauder, was introduced, and it carried the high-output V-8 powerplants. The FE family of engines was long-lived and used in a wide variety of guises. From grocery-getter to race car, this group of engines was used through 1971 and could be relied on to produce serious power. It was an extended skirt– style block, like the old Y-block. It had thick-wall cylinders and the entire engine tipped the scales at over 700 pounds. When Ford stepped away from its self-imposed ban on racing in 1961, it had a powerplant that could, and did, produce the results that put it in the winner's circle.

The 8.9:1 compression and a two-barrel carburetor 352-ci engine was called the "Marauder 352," and it put out 220 horsepower. The "Marauder 390" was the next engine higher in the food chain. With a 9.6:1 compression ratio and a four-barrel carburetor, it made 300 horsepower. Later in the 1961 model year, another "Marauder 390" designed for use in police vehicles, was released. It was rated at 330 horses, replete with "Police Special" valve-cover decals. In the March 1961 issue, *Motor Trend* tested a 300-horsepower 1961 Marauder that was equipped with a Multi-Range Merc-O-Matic transmission and 3.00:1 axle ratio. The four-door got up to 60 miles per hour from rest in 10.2 seconds, not bad numbers for a vehicle weighing 3,795 pounds.

By model year 1962, things were warming up at FoMoCo, especially in the engine room. The Marauder family of engines was expanding, and the results were highly impressive. A new showcase for the big-blocks appeared—the mind-melting Monterey S-55. With a choice of convertible or hardtop, bucket seats, and either an automatic transmission or four-speed manual, the S-55 served notice that the traditional, stodgy Mercury image was going to be blown into the weeds. In midyear, a new Marauder 406 V-8 was released, and it boasted 385 horsepower with a four-barrel carburetor. With the same 10.9:1 compression ratio, a trio of two-barrel carbs, the 406 V-8,

A chrome-edged belt line was a typical styling cue of the day, as was an ornate, difficult-to-clean grille.

This car came with a $379.90 option that was essential for extreme high-performance use: cross-bolted main bearing caps. This slugger had the cross-bolted main bearing caps on the number 2, 3, and 4 main bearings, which prevented cracking of the block. If the center bolts on those bearing caps loosened under the stress of high-speed operation, the crankshaft was permitted to flex, and that risked engine failure. The buyer could only order this engine with cross-bolted mains if the following items were left off of the car: air conditioning, power steering, power brakes, and automatic transmission.

In testing, the *Motor Trend* crew wound it up and came away with a 0-60 time of only 7.65 seconds. The Marauder covered the quarter-mile in 16.5 seconds, tripping the lights at 94 miles per hour. Keeping the accelerator planted resulted in a top speed of 120 miles per hour. These were numbers people did not normally associate with Mercury, but the coming years would show that this was not a fluke. The performance profile of the Mercury Man was going to get higher, very quickly.

Marauder, Two Tons of V-8 High Performance

Sizzler. That's how Mercury announced its hot, go-fast machine for 1963. The Marauder was introduced as a midyear release, and it was a time of real performance gains. Mercury was making inroads into NASCAR and USAC (United States Auto Club), and the publicity had the desired effect in the showroom. The sanctioning bodies required the manufacturers to race products the public could buy over the counter. As a result, the ordinary buyer was able to stuff the family garage with a race car fitted with a license plate frame. The buyer just had to find a way through the labyrinth of options.

The Monterey line in 1963, for example, comprised four series with seven body styles and 16 different models. They all rode on the same 120-inch wheelbase and measured 215 inches in length. The Ford/Mercury "Breezeway" rear window, as seen on the 1957-58 Turnpike Cruiser, was reintroduced. In a straight line, the sedans held their own with the competition of the day. But from a performance standpoint, the Marauder S-55 was a real musclecar that deserved scrutiny. This model had a new "fastback" styling, introduced on the Marauder Convertible show car that made the auto shows in early 1963. The NASCAR circuit would know this competitive, aerodynamic model. The standard engine was the Marauder Super 390 V-8, which boasted of 300 horsepower at 4,600 rpm and 427 foot-pounds of tire-melting torque at 2,800 rpm. The 10.8:1 compression ratio required

became the "Super Marauder," kicking out 405 horsepower at 5,000 rpm. With two-bolt main bearing caps, cast-iron crankshafts and mechanical valve lifters, these engines had little trouble turning the tires of the day into rubber dust. *Motor Trend*'s May 1962 issue put a Monterey Custom convertible, equipped with the 300-horsepower "Marauder 390" engine, through its paces. At the end of the long straight at Riverside Raceway, the magazine's Weston electric speedometer read 110 miles per hour, and the Mercury felt like it had a few more miles per hour under the hood. With a 0-60 time of 10.5 seconds and a quarter-mile time of 18.9 seconds, the Multi-Range Merc-O-Matic Marauder 390 nevertheless paled in comparison with the S-55 that *Motor Trend* would test in its October 1962 issue.

The Marauder convertible was home to a "Super Marauder 406," which belted out 405 horsepower.

premium fuel flowing through the four-barrel carburetor, and the power could be fed through either a four-speed manual floorshift transmission or the Multi-Drive Merc-O-Matic automatic. Mercury's enthusiasm and excitement for the car were readily apparent. "Shift to the real performer! Go Mercury!" read a period ad.

If the buyer wanted a bit more, the 406-ci "Police Special" was back. Its 600 cubic feet per minute (cfm) four-barrel carb and 10.5:1 compression helped produce 330 horsepower at 5,000 rpm and 427 foot-pounds at 3,200 rpm. Early in the model year, the top engine options were the four-barrel 385-horsepower Marauder 406 V-8 and the six-barrel Marauder Super 406 with 405 horsepower at 5,800 rpm. The 405-horsepower 406 had a trio of "progressive" mechanically linked Holley two-barrel carburetors on an aluminum two-plane intake manifold that allowed 900 cfm of gas-air mixture to flow into the cylinders. The "406" engine was only offered through spring of 1963, when a pair of new engines replaced them—the famed 427s.

A Marauder on Steroids, the S-55

The Ford and Mercury engine development program shifted into high gear during the mid-1960s. Lee Iacoccoa led Ford's emphasis on high performance. Ford had built an experimental 483-ci engine that pushed a 1962 Galaxie to 172.26 miles per hour at the Bonneville salt flats, but NASCAR and the NHRA (National Hot Rod Association) put a 7-liter (428 ci) limit on engine size, so the larger engine was set aside. Development of the 406 engine resulted in the 427 powerplant. Introduced as a midyear pair of options, the 427 Super Marauder V-8s were racing engines that were dropped into a street car. The standard or base 427 produced 410 horsepower at 5,600 rpm with torque coming in at 476 foot-pounds at 3,400 rpm. It breathed through a 780-cfm Holley carburetor atop an aluminum intake manifold, while the mechanical valvetrain regulated the mixture into the combustion chambers. If this was not enough power, a buyer would check the "427 8V Hi-Perf" option box for an extra $461.60 and take home a genuine performance leader. The engine was equipped with a plethora of high-performance features, such as dual quad 780-cfm Holley carburetors on an aluminum intake, cross-bolted bearing caps, and 11.5:1 compression ratio that demanded a Super Premium fuel. This top level engine was rated at 425 horsepower at 6,000 rpm; it would rev to 7,000 rpm without much trouble, and its 480 foot-pounds of torque made tire salesmen very happy.

Triple taillights and tiny fins were part of Mercury's upscale styling. The vertical-ribbed grille, front and rear, has been a Mercury tradition since the division's birth.

In the April 1963 issue of *Car Life*, a Mercury "427 Super Marauder" S-55 Hardtop was put to the test. This was a notchback model, as the term Marauder applied to the engine in this instance. This engine was available in most of the large Mercury lineup, but the transmission choices were limited to the four-speed manual. The numbers that the drivetrain pulled were indicative of the grunt under the hood. From 0 to 60 miles per hour took 7 seconds, the quarter-mile disappeared behind the car in 15.1 seconds at 87 miles per hour, not bad for a vehicle that weighed about 3,900 pounds. *Car Life* complained that the 8.00x14 tires didn't grip very well under full-throttle acceleration. Really! This from a car that literally blew its mufflers off the header pipes after several high-rpm runs. They must have forgotten that this was a racing engine,

The 1963 Mercury Meteor was an intermediate-sized car that was dropped from the Mercury lineup in 1964, because it was a direct competitor to stable-mate Comet.

placed in a passenger car to satisfy the requirements of the race sanctioning groups.

Racing was definitely in Mercury's mind. Although Mercury's motorsports involvement is covered in detail in another chapter, it must be pointed out that racing did help market cars. Big Mercurys sold at an increasing rate, 121,048 units for model year 1963, the best showing since 1960. In an advertisement from 1963, Mercury used race results to "establish itself as the *new* performance champion of the medium-price field." This showed the suits at Mercury that the Comet needed to have a competition back-

ground to increase its public visibility. The full-sized Mercurys were doing well in NASCAR, and though the division was backing racing via "indirect" support, it used a safety angle to rationalize its involvement. Ben Mills, Mercury's division manager, put Mercury's racing activities into the proper perspective. "The racetrack is a logical extension of our engineering programs. Cars with 110 horsepower or less have two to three times as many accidents, per mile, as cars with 300 horsepower." The statement explains Mercury's rationale for offering the public an engine with 425 horsepower. However, Ford Motor Company realized that this was not your typical engine offering. The conventional 24-month/24,000-mile warranty was tossed out and replaced with a 3-month/4,000-mile power-train warranty. Over its lifetime, the 427 proved its worth on a wide spectrum of racing venues from stock cars to drag racing to LeMans. The "FE" engine had grown up.

The Marauder Gets a Facelift

The major differences between the 1963 and 1964 models were the trim. Trim pieces were changed, but nothing major, and the S-55 was dropped. Mercury felt that sales were going well, and why mess with success? Lincoln styling influence was exhibited on the large Mercurys, especially the rear, and the 1963 Thunderbird seemed to be the source of '64 full-size Mercury front ends. The Marauder name now graced the convertibles, the two- and four-door pillarless fastback design of both the Monterey and Montclair lines, and the resurrected top-line Park Lane. *Car Life* magazine did not speak highly of the Park Lane, calling it a warmed-over Ford on the outside and a gaudy car on the inside. This was a full-tilt luxury car that could also be equipped, like any other full-sized Mercury, with the 427-ci engine. Testing a Park Lane resulted in a 0-60 time of 9.3 seconds. Its 4,050 pounds went down the drag strip in 16.9 seconds tripping the lights at 83 miles per hour. The magazine topped out the vehicle at 112 miles per hour, but still said that "it fails to meet its heritage of 25 years as a luxury performance car." It wasn't an accurate statementænot many Mercurys had posted numbers like that in the preceding 25 years.

In the Marauders, the base engine was the two-barrel 390-ci, 250 horsepower, 378-foot-pounds V-8. A two-barrel carburetor, 266-horsepower version was added for use with automatic transmissions. The 300-horsepower 390 was next on the option sheet; equipped with a four-barrel carb, 10.1:1 compression ratio, and hydraulic lifters, the powerplant attained maximum horsepower at 4,600 rpm. The torque rating was unchanged for 1964 with 427 foot-pounds at 2,800 rpm.

The "Police Interceptor" engine moved on to the regular option list. It boasted a solid-lifter camshaft, 10.5:1 compression and 330 horsepower at 5,000 rpm, and abundant stump-pulling power with 427 foot-pounds of grunt at 3,200 rpm. If a buyer "needed" more of everything, the dynamic duo were back, unchanged. The 427-ci "7-Liter" engines remained the same in 1963, just as potent and just as fearsome. The Mercury copywriters had fun creating ad copy for this powerplant, calling it "the engine that

set a new world's stock-car record in the most recent Pikes Peak Climb." This was in reference to Parnelli Jones being the first to the top of the Colorado mountain in a Bill Stroppe-prepared Mercury Marauder in 1963 and 1964. The 410- and 425-horsepower versions of the big-block were still warming tires on the street and on the track.

Mercury's Lincolns—Monterey, Montclair, and Park Lane

By the mid-1960s, Mercury had taken on new and distinctively different styling. The slab-sided body style of the Lincoln division found its way into the Mercury styling studio, resulting in a completely new appearance for the full-sized offerings from Mercury in 1965. Advertisements of the day boasted that the restyled Mercurys were "Now in the Lincoln Continental Tradition!" The wheelbase was lengthened 3 inches to 123 inches. The Breezeway sedan was still in production, and it retained the slanting and retractable rear window. The pillarless hardtops in the Monterey, Montclair, and Park Lane series were called Marauders, just like in 1964, even if they didn't look like the preceding year's cars. From the upright grille to the vertical taillights, it looked like its upscale divisional partner. It was a massive yet handsome car with an option list as long as your arm. It wasn't difficult for the vehicle weight to exceed two tons. What better way to get that mass in motion than with a big-block engine?

Mercury did not revise its powerplant lineup for 1965. Rather, they spent no small amount of time shaping the new sheet metal. Again, the two-barrel 390-ci engine was the base offering. Like the prior year, the power outputs were stepped to allow a buyer to choose just the right amount of thrust. And also like the prior year, the biggest of the FoMoCo big-blocks could live under the hood of a big Mercury. For a little more cash, the buyer could get a lot of dash. For $340.90 one could buy the 410-horsepower 427 mill in the Park Lane, and $388.70 secured the engine for the other Mercurys. And dual four-barrels for the 427 were still an option. Of course, a four-speed manual transmission was mandatory with either 427 engine.

Little did the public know that this was the last year for this beast in a full-sized street Mercury. There was definite market shift away from full-size, heavyweight, big-inch performance to compact, lighter weight cars with improved power-to-weight ratios. The competition was putting its performance dollars in smaller platforms, such as the Pontiac Tempest LeMans GTO equipped with a 389-ci engine. While

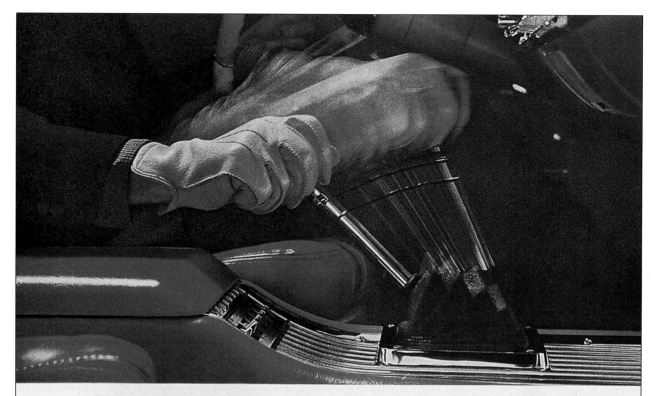

Shift to the real performer! Go Mercury!

Monterey S-55 2-door hardtop

Feel the brilliant performance that is now standard in every Mercury—a big Marauder 390 V-8. Team up this engine with Mercury's optional fully synchronized "4-on-the-floor" stick-shift transmission and there's a new sense of oneness between you and your car. Effortless cornering, passing, and mountaineering. Want the best? Stop in at Quality Headquarters—your Mercury dealer's.

MERCURY
MONTEREY·MONTEREY CUSTOM
MARAUDER and S-55

COMET·METEOR·MERCURY...PRODUCTS OF *Ford* MOTOR COMPANY...LINCOLN-MERCURY DIVISION

this powerplant was little threat to the 427s, the vehicle weight was considerably less. Performance was enhanced, it cost less money to build and buy, and the public snapped them up. The crew at Mercury was aware of all this, and had their reply in the pipeline. The only trouble was, it was a fairly long pipe.

A Revamped and Revitalized S-55

In 1965, Mercury underwent several significant changes: the "sporty" S-55 option (center console, bucket seats, and dual exhausts) returned, available on either the convertible or the two-door hardtop coupe; the Marauder model was dropped from the Mercury ranks, but the name lived on in the engine bay; and the fearsome 427 engine disappeared as well. The power source under the hood was the Super Marauder 428-ci V-8, which was far different from a 427. The stroke of the 428 was shorter than the 427 while the 428's bore was larger. This "squarer" configuration was a more sedate version of the big-block FE engine. It produced enough power to avoid apologies, but it came on the heels of the 427, which was a tough act to follow. The 428 had the same bore as the 406, but it did not enjoy the same superior oiling system as that engine. The lubrication configuration of the 428 was the same as in the standard 390; that is, a restriction was designed into the system to force oil into the lifter galleries, and the main oil galleries are of a smaller diameter. With very thin cylinder

The 1964 Marauder showed off a pillar-less design and fastback styling. The swept rear window treatment was developed for use on NASCAR racetracks. *Ford Motor Company*

The Marauder X-100's hidden headlights were a touch borrowed from the Cougar, resulting in a sleek look, as long as the headlamp doors were shut. Getting the doors to line up when shut was always a challenge, and being a production line product, occasionally a misalignment would occur.

walls, boring out the engine over .030 inches was not recommended. Unlike earlier FE engines, the 428 was fitted with an external balancer. Two-bolt main bearing caps holding cast crankshafts were the rule when the engine was introduced in the 1966 S-55 Mercury. Other internal details included a hydraulic cam, shaft-mounted rocker arms, a dual-plane intake manifold, and cast-aluminum pistons. With a Ford C6AF-9510-AD four-barrel carburetor and a 10.5:1 compression ratio, the cast-iron V-8 put out 345 horsepower at 4,600 rpm, and it created 462 foot-pounds of torque

at 2,800 rpm. These numbers were nothing to be ashamed of, but the industry had entered the muscle-car wars and a two-door hardtop weighing 4,031 pounds wasn't breaking new ground.

Fitted with either Multi-Drive Merc-O-Matic or the top-loader four-speed manual, the 1966 S-55 put more sport into full-sized Mercury than any other offerings in the Lincoln-Mercury family. But sport to a street Mercury was not the same as the competition. *Motor Trend* (August 1966) road-tested a two-door hardtop S-55 that was abundantly optioned at 4,260

Simulated exhaust ports helped break up the vastness of the side, while the rear-wheel opening cover imparted a Lincoln-like visual style.

pounds. While not a nimble package, *MT* praised the Mercury's cornering prowess. The 51/49 front-to-rear weight-distribution ratio did not hurt ad claims. Equipped with the base "428" engine, developing 345 horsepower, the vehicle had impressive acceleration numbers: 0-60 in 8.9 seconds. It ran the quarter-mile in 16.9 seconds at 85 miles per hour. The magazine noted that even with its considerable power, "its full impact isn't felt until after some momentum is built up

to overcome the weight of the car." The optional front disc brakes needed 169 feet to haul the S-55 to a halt from 60 miles per hour, and with an average fuel consumption of 12 miles per gallon, the unusual vertical 25- gallon gas tank allowed for more trunk room. *Motor Trend* liked the easy-to-clean synthetic seat covering, but not the fact that a healthy dose of options drove the sticker price of the S-55 higher than $5,000. Just 2,916 S-55 hardtops and only 669 S-55 convertibles were sold in the 1966 model year.

Cougar, Muscle with Incomparable Style

Full-sized, high-powered "sport" vehicles were falling into disfavor with the buying public. The smaller midsize or compact "Musclecar," such as the Ford Mustang, Chevy Camaro, and Pontiac GTO, took people from Point A to Point B as quickly as possible, while preferably humiliating another rival musclecar. Mercury buyers interested in a high-performance sports car bought the new 1967 Cougar. When the 1967 Mercurys rolled into showrooms, they had massaged sheet metal again. The slab-sided formal crispness of the '66s gave way to a more flowing look, especially in the two-door hardtop with its concave rear window. Admittedly, the front end did have a more pronounced Lincoln look, and the famed Breezeway hung on with a forward slanting rear window that lowered only 2 inches. Due to required safety equipment as well as economic inflation, retail pricing rose, so it came as no surprise when fewer S-55s were sold in model year 1967. Offering the S-55

The Marauder X-100 was only offered two years, but the matte finish on the rear deck was available only in 1969. While this car provided adequate straightline performance, it was barge-like and handled like other luxury car offerings of the day. *Ford Motor Company*

The 429-ci engine had little trouble filling the Marauder X-100 engine compartment. This large engine was required to drive innumerable power options as well as propel the car forward in a commanding fashion.

as a Sports Package option in the Monterey line, Mercury built 570 hardtop cars and only 145 S-55 convertibles. Front disc brakes, engine chrome, and additional sound insulation (weight) were standard on S-55s. As in the year before, the 10.5:1 compression Marauder 428-ci engine with an output of 345 horsepower at 4,600 rpm was standard, as was a torque rating of 462 foot-pounds at a leisurely 2,800 rpm. These cars were low-rpm torquers, not high-rpm screamers with blazing acceleration and amazing speed. Tipping the scales at 3,956 pounds for the hardtop and 4,093 pounds for the convertible ensured that road imperfections were a nonevent. The S-55 badge whispered promises of performance, but the marketplace deserted it. And when that happened, Mercury let the S-55 fade away with the end of the 1967 model year. The full-sized performance Mercury became a memory.

X-100, Big Inch Luxury

In model year 1968, Mercury was busy building all of the hot-selling Cougars it could produce, the new Montego and Cyclone lines expanded, and the Cyclone stepped into the role of racing flagbearer, a spot that the Marauder had enjoyed in 1963-64. Mercury's plate was full. Lee A. Iacocca, Ford's group vice president of Lincoln-Mercury, was guiding the company to an exciting high-performance future. After the overwhelming success of the Mustang , in January

1965 Henry Ford II had tapped Iacocca to lead Lincoln-Mercury. "We've been at this 20 years and we're not making it, so let's give it one last college try," Ford had said. His goal was to allow Mercury to claim a larger piece of the musclecar pie. Ford wanted some of the success that Iacocca had achieved with the Mustang to rub off onto the division that many saw as a step-child, and Iacocca had definite ideas about raising the visibility of Mercury by tying it more closely with Lincoln, specifically with the full-sized automobiles.

In 1969, the year of Neil Armstrong's "stroll" on the moon , Mercury re-released a name that had written high-performance history for the company. The division was hoping to capitalize on the Marauder and Marauder X-100. Now it graced the body of what was basically a top-of-the-line Mercury Marquis in a two-door hardtop configuration. Lee Iacocca said in *Motor Trend* in September 1968 that "We're trying to create a luxurious identity for the car line. We want a plush feel. Take the Marauder. It's smaller than the Marquis. But look at it—when I see an X-100 Marauder all jazzed up you think that its got the Marquis Lincoln type front end but I read it as an overall sporty car—but with a luxury look." He tied the division knot tighter with "the Marauder by Lincoln-Mercury."

Marauder Styling Elegance

A. B. (Buzz) Grissinger, the director of Lincoln-Mercury styling, spoke to *Motor Trend* for its September

The X-100's Rim Blow steering wheel was a questionable feature, and it was often triggered accidentally when the car was being parked. A shortage of instruments meant that the driver was kept in the dark regarding what was going on under the huge hood.

In 1963, Mercury wasted little time telling the public that the Marauder was related to these race winners. Since Mercury had a very conservative reputation, these advertisements helped establish the Ford "Total Performance" image. *Ford Motor Company*

MERCURY announces its newest sizzler, the...

Marauder

Even the styling says "go." Note that slim, racy new hardtop roof. It's not only beautiful, its aerodynamic styling reduces air resistance. Choose from two dashing models: the Marauder with a big 390 V-8 standard...or the sporty Marauder S-55, with console-mounted transmission selector, luxurious bucket seats, and a 4-barrel Super Marauder 390 V-8 standard engine. Options range up to a high-performance Super Marauder 427 V-8. Talk about hot! Talk to your Mercury dealer.

MERCURY
MONTEREY·MONTEREY CUSTOM
MARAUDER and S 55

COMET · METEOR · MERCURY: PRODUCTS OF ⬭Ford⬭ MOTOR COMPANY · LINCOLN-MERCURY DIVISION

With its new fastback styling, the 1963 Marauder was the forerunner of the musclecar. Its sloped rear window was a benefit to racers, and a 427 side-oiler could be slipped under the hood. *Ford Motor Company*

1968 issue and profiled the 1969 models. "We've tried to carry forward the association with the Lincoln-Continental—the appearance, dignity, and elegance. The Marauder retains all the features of the Marquis but has some design changes to convey the 'King of Speed' feeling." A two-tone paint job was the most obvious difference between the Marauder X-100 and the rest of the upscale Mercury line. The rear deck behind the back window was recessed and painted a contrasting dark matte color. The tunnel effect swept the trunk lid down and over the taillights. The rear wheels of the X-100 were covered with fender skirts (optional on the Marauder) and a false air intake was behind the doors on the rear flanks. While not quite as large as a Marquis' 124-inch span between wheel centers, the X-100 rode on a 121-inch wheelbase to the tune of 4,009 pounds before the options were ladled on. The same full-sized approach was found inside where a leather and vinyl interior, electric clock, and a rim-blow steering wheel helped to convey the "master of all he surveys" feeling when gazing over the long hood of a Marauder X-100. It was actually a well-proportioned, graceful-looking vehicle in the American massive idiom.

When you are driving the "King of Speed," it's necessary to have royal power under the hood. Mercury fitted both of its entries to the luxury car/high-performance field with engines up to the task. The Marauder was equipped with the venerable "FE" block, 390-ci mill. Its 9.5:1 compression and two-barrel carburetor produced 265 horsepower at 4,400 rpm. More importantly, the car with a curb weight of 4,500 pounds fully loaded cranked out 390 foot-pounds at 2,600 rpm. But for those who wanted to stand out from the crowd, the X-100 was the choice. Lurking under the huge hood was a huge engine—the 429. The 429, part of the "385" family, was introduced in 1968 and designed for use in "big luxury cars." It was used in Lincolns in a 460-ci application. This was a 429 bored out to 460. The engine's thin wall casting, huge bearing surfaces, and heads showcased the latest in "Poly-Angle" valve design with a canted-valve cylinder head arrangement. The staggered valves configuration had each rocker arm riding on a fulcrum on a separate pedestal and not on a rocker shaft. The racing design influence helped the behemoth to show 360 horsepower at 4,600 rpm with a four-barrel carb resting on top. The 10.5:1 compression helped produce 480 foot-pounds of torque at 2,800 rpm. This hydraulic lifter–equipped engine was designed with forthcoming emission regulations in mind, as well as propelling a Mercury down the road in a commanding fashion.

During road tests of the day, it was discovered that it was possible to transport considerable weight at serious velocities, if the car was equipped with a big-inch engine. *Car Life* timed the X-100 at 7.5 seconds on a 0-60 miles per hour test, while another magazine, using an X-100 fitted with a highway-oriented 2.80:1 rear-axle ratio went from 0 to 60 in 8 seconds flat, covering the quarter-mile in a tick under 16 seconds at 86 miles per hour, and topping out at around 125 miles per hour. Front disc brakes provided an added and necessary measure of stopping power. Large steering inputs at speed were discouraged. With four turns lock-to-lock, the driver felt like the helmsman on an ocean liner, experiencing vague steering feedback and a large amount of body roll.

The Marauder base model sold 9,031 units, and the two-tone X-100 sold only 5,635. Mercury went into the 1970 model year hoping for an increase in sales, but that hope was dashed. Disappointment was around the corner for Mercury management. Sales figures were even lower than the dismal 1969 numbers, and the base Marauder sold 3,397 while only 2,646 X-100s were sold, despite some discounting and a price cut. When Mercury brass saw those sales figures, the plug was pulled. After this, Mercury's performance efforts were concentrated on the mid-size lines, but the growing emission regulations would soon reduce these proud cars to a shadow of their former selves. At least the Marauder and Marauder X-100 went out strong, not reduced to an embarrassing caricature of a performance vehicle.

Improving the Species

Mercury's Racing Program

4

When English naturalist Charles Darwin wrote his theory of natural selection, he did not have automotive competition in mind, yet the arena of motorsports is a good example of the survival of the fittest. By the 1950s, Mercury was heavily involved in racing. The division supplied the 1950 Indianapolis 500 with a pace car, and two years later Clay Smith and Troy Ruttman drove a 1948 Mercury Coupe to fourth place in the 1952 Mexican Road Race. But that independent effort had little effect on the division's interest in sanctioned competition. The very first NASCAR race in June 1949 was won by Jim Roper's Lincoln, but the sister company, Mercury, had little involvement in motorsports during the late 1940s. As the 1950s progressed, Mercury gained a higher profile on the racing circuits. Merc racers started to become a familiar sight on the starting grid. They weren't always winners, but they learned valuable lessons that would help ensure the survival of the species.

Mercury Goes NASCAR Racing

When "Big" Bill France started NASCAR in 1947, he had high hopes that it would someday become the most successful racing series in U.S. history. In the 1930s France had built and raced Fords, and in 1936 entered a race in his hometown of Daytona, Florida. Sir Malcolm Campbell raced his Bluebird land speed race car on the area's wide, flat beach. Eventually, the Briton moved his the land speed efforts to the Bonneville Salt Flats, but city officials wanted to retain their racing reputation. So the

"Dyno" Don Nicholson started racing this Mercury Comet station wagon while his teammates, driving coupes, complained that Nicholson had an unfair advantage due to the wagon's shorter wheelbase and weight in the rear. "Dyno" switched to a coupe to appease his teammates, and he still won with the coupes. Here Nicholson drives through the parking lot at Detroit Dragway in the spring of 1964; it looks cold. *James J. Genat/ Zone Five Photo*

Darel Dieringer stands with his 1963 Mercury Marauder that is equipped with a 427-ci side-oiler engine. *Daytona Racing Archives*

Daytona city fathers hosted a 3.2-mile race on the beach and adjacent road. Sanctioned by the American Automobile Association (AAA), the 250-mile oval pattern race was ended after 75 of the planned 78 laps due to the arrival of the pesky tide. France saw that the public enjoyed the spectacle and the sport could be lucrative. He promoted a number of stock car races in the late 1930s and in the late 1940s after World War II. Without official sanctioning, a champion couldn't be crowned. Hence, a group of independent racers, promoters, and track owners, including France, met and formed NASCAR. The fledgling organization held its first race at a 3/4-mile dirt track near Charlotte, North Carolina. Glen Dunaway, driving a Ford, won the race, but he was stripped of the win when officials found illegal modifications to the car. The win was awarded to Jim Roper, driving a Lincoln.

Race winner Cale Yarborough, in the #21 Mercury Cyclone, battles Lee Roy Yarbrough, in the #26 Mercury Cyclone on the high banking during the 1968 Daytona 500. The Mercury Cyclones had a unique aerodynamic fascia that gave them a competitive advantage. *Daytona Racing Archives*

The famed 427-ci engine. Known popularly as the "side-oiler," it earned a place in the heart of racers for many years. Ford relentlessly campaigned to have the engine homologated for NASCAR competition, but in the end it was to no avail. *Ford Motor Company*

From these beginnings, NASCAR grew and sanctioned races throughout the Southeast. Racing on Daytona Beach was spectacular; sand flew as the full-sized stockers drifted through the corners like over-sized sprint cars. With the prospect of winning on Sunday and selling on Monday, Mercury was in the thick of it. Tim Flock's M355 was one of a trio of Mercury's built by Bill Stroppe. Stroppe had been recruited by Mercury to develop race cars to counter the Ford contingent. Equipped with a 368-ci Y- block engine, the M355, the 335-horsepower race car started life as a production vehicle, but it spent a fair amount of time being modified. The suspension was substantially reinforced, and each wheel utilized two shock absorbers to withstand the rigors of racing. The powerplant under the hood was far more interesting than the ventilated drum brakes that helped slow the race car down. From its dual four-barrel carburetors atop an aluminum intake manifold to the high-compression pistons and forged steel truck crankshaft, this race car had a top speed in the neighborhood of 117 miles per hour.

FoMoCo was on a victory roll in early 1957. By mid-season, Ford had won 15 Grand National races to Chevrolet's five. In February of 1957, the infamous Automobile Manufacturers Association (AMA) ban was proposed by General Motors, and the way the game was played changed. Under terms of the agreement, all of the manufacturers agreed to end factory-sponsored racing and remove all of their high-performance equipment from their catalogs.

Racing was not to be used as a medium for marketing, sales, or promoting safety and style. While Ford Vice President and General Manager Robert McNamara followed the ban, General Motors ignored it and flourished on the track. Yet the Stroppe-prepared Mercurys, driven by Ruttman, Sam Hanks, Jimmy Bryan, and Billy Myers, were winning races for an under-financed division. As Stroppe put it, "There was no love lost between Ford and Lincoln-Mercury divisions. They were run like separate companies."

After the ban went into effect, Ford won 12 races and Chevrolet took home 14 checkered flags. Ford production car sales in 1958 were in the tank, and the bean-counters were sure that severing ties with racing was saving the company money. GM cleaned up on the track and in the showrooms. The winged 1959 Chevrolets racers that enjoyed success on the track were a sales flop. Ford Motor Company sales increased over 500,000 units sold from 1958.

McNamara allowed some funds to trickle their way into competition, and Ford entries operated very well on a shoe-string budget. Ford developed a new production engine that it could slip into stock car racing—the "FE" block 390-ci mill. But the suits would not risk violating the AMA ban, so this engine was put on hold. FoMoCo relented a bit in the pursuit of performance-parts development by forming a group to test the waters of a "limited re-entry" into racing. NASCAR continued to expand and develop as a series and several new superspeedways opened in 1960. Ford wanted to increase its profile in racing

Crew members scramble as "The Silver Fox," David Pearson, prepares to roll out of pit lane in the Wood Brothers #21 Purolator 1978 Mercury Cyclone at Talladega. Pearson drove to a solid fifth place finish in the race. *Daytona Racing Archives*

Curtis "Crawfish" Crider held the track record at Darlington in 1961 (119.854 miles per hour) behind the wheel of a Mercury. Racing from 1950 to 1974, he was one of the many good drivers that never stood in the spotlight, but he fielded fast cars. *Curtis Crider Collection*

and, thus, gain a larger share of the sales pie. At the corporate level, the spirit was willing, but the top honchos were still not interested to pursue racing as the competition did. Finally, fate intervened, when President Kennedy appointed Robert McNamara as Secretary of Defense. This opened the door for a man who would become a household name—Lee Iacocca. The entire motorsports policy changed, and the floodgates of money needed to win opened. Iacocca was the right man, in the right place, at the right time, and at age 36 he became the youngest general manager of Ford. And he saw to it that the Blue Oval and its sister divisions reaped the successes of racing. What he had in mind was nothing short of total performance.

Mercury's Top-Line Racing Effort

Success in racing does not happen overnight. Careful planning, good people, and adequate funding

are essential for any long-term racing program to deliver the desired results. Ford's plan was to hang on to the few teams still using Ford products until better weapons could be developed and put into their hands. Mercury was eager to field a race team. The adage, "win on Sunday and sell on Monday," proved correct many times over.

Ironically, General Motors decided to terminate all factory-backed motorsports involvement in early 1963, due to the possibility of anti-trust legislation. General Motors was, at the time, the largest automotive manufacturer in the world, and the executives wanted to lower the corporation's profile. All of the wins by GM directly and indirectly backed teams were getting "The General" too much attention from Washington, D.C.

Mercury did not want to appear to have a direct factory involvement in racing, but in 1963, Stroppe was contacted and told to put a team together. He would be responsible for the team's success and would take any knocks for Mercury if the marque didn't find victory lane. The famed drive-anything racer Parnelli Jones secured the win at Pikes Peak for Mercury in a "Super Marauder," equipped with a 427 side-oiler engine. Before joining Stroppe's team, Jones had won the 1963 Indianapolis 500 and was that year's USAC National Champion. He hit the United States Automobile Club circuit like a whirlwind, winning eight major stock car races in a Mercury Marauder. In 1964, Jones was back again for a return visit to "The Hill." And though it wasn't his favorite race, Jones scored a repeat win at the Pikes Peak Hill Climb.

In 1964, a fleet of red, white, and blue Marauders prepared by Stroppe romped to victory on asphalt ovals and road courses. The surprisingly stock 3,715-pound race cars, powered by the 427s, achieved speeds around 175 miles per hour at the Daytona superspeedway through tuning, tweaking, and attention to detail. Darel Dieringer took home the winning trophy at the NASCAR race held at Augusta, Georgia. At Riverside Raceway in January 1964, one of stock car's racing legends lost his life, 1962/63 Grand National champion Joe Weatherly, driving a Bud Moore Marauder, hit the Turn 6 wall. Billy Wade took over Weatherly's seat and put the big Mercury in the win column four times. Tragically, Wade was killed when his Mercury made heavy contact with the Turn 1 wall during tire testing at Daytona in January 1965. Mercury earned five wins by season's end and announced its withdrawal from NASCAR. Why? Mercury never made its reasons known, but Stroppe was told to wind down his operations. Stroppe, who got

A PRODUCT OF (Ford) MOTOR COMPANY · LINCOLN-MERCURY DIVISION

We'd like to keep you posted on Comet's wins ...but they're mounting up too fast

Part of what Comet did within a two-week period

June 5	Cecil County Drag-way, Maryland	Match Race	J. Chrisman	Won 2-of-3 match in 2 straight. Best ET: 9.60
June 4-6	NHRA Spring-national, Bristol, Tenn.	A/FX	D. Nicholson	Class winner with low A/FX ET of 10.52
May 30	U.S. 30 Dragway—Gary, Indiana	S/S	E. Schartman	S/S Eliminator over field including "funny cars." Best ET: 10.71
May 23	Long Beach Drag-way, Calif.	Match Race	H. Proffitt	Set new strip ET mark of 10.53, taking 2-of-3 match race

The record proves that competition-modified Comets are consistent winners, not once-in-a-while luck cars. So if you like performance, visit your Mercury dealer. You'll see that the whole Comet line has winning ways. The new Cyclone 2-door hardtop is a prime example. It comes with a 4-barrel Cyclone Super 289 V-8 that checks in at a hefty 225 hp. With bore and stroke at 4.00x2.87, and compression ratio at 10.0, this mill is designed to earn its name. 4-on-the-floor or Merc-O-Matic optional; a full-synchro 3-speed manual, standard. Buckets and console, tach, engine dress-up kit and chromed, lugged wheel covers all included. Wouldn't you like to be included too? Your Mercury dealer can fix it up.

 *Mercury* Comet

the world's 100,000-mile Durability Champion

 RIDE WALT DISNEY'S MAGIC SKYWAY AT THE FORD MOTOR COMPANY PAVILION, NEW YORK WORLD'S FAIR

Mercury used the exploits of the factory drag team in 1965-period advertisements. Mercury was racking up an impressive number of wins on the drag strip. *Ford Motor Company*

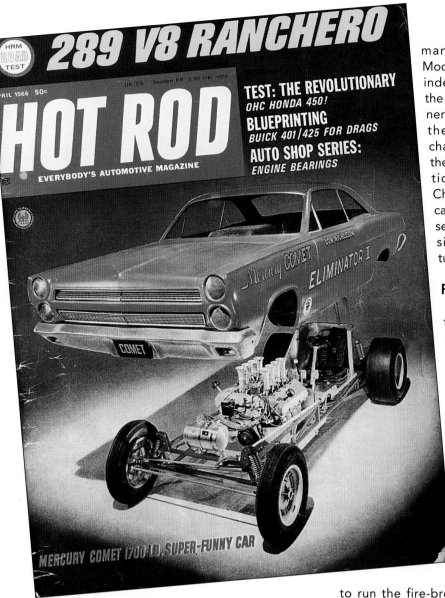

Hot Rod magazine featured Nicholson's new "Funny Car" on its April 1966 cover. Nicholson brought Mercury performance to the forefront of drag racing and beat many Hemi-powered Mopars during the 1960s. *Petersen Publishing*

the word to shut down his USAC stock car setup as well, got the factory to let him hold onto the equipment until the end of the season. The factory agreed. With the heroic and talented Parnelli Jones at the wheel of his race cars, the team grabbed seven victories over the course of the season. It was a banner year for Mercury racing. Stroppe won the owner's title, and Mercury won the manufacturer's championship There's nothing like leaving on a high note!

In 1965, the Ford-built Galaxie enjoyed success on the NASCAR circuit. Meanwhile, Dieringer started

many races in a one-year-old Bud Moore–prepared Marauder that was an independent entrant. Unfortunately, the car did not find its way to the winner's circle. Without factory backing, the big Mercurys didn't stand a chance of defeating the Fords. And there wasn't really any other competition, as a regulations battle left Chrysler out of contention. Ford race cars won forty-eight races out of the season's fifty-five events. Rather lopsided. However, the tables were turned the next year.

Ford's NASCAR Follies

Ford had tried and tried again to legalize its incredible 427 single overhead cam (SOHC) racing engine. This powerplant was a terror to the competition on the drag strip and Dearborn wanted to put it in the NASCAR arena to duplicate its 1965 success streak. Initially, NASCAR ruled that it wasn't a production engine. Then Chrysler started selling street versions of its famed Hemi, which qualified the engine for NASCAR racing. In light of the Hemi engine homologation, Ford appealed to the Auto Competition Committee (ACCUS). Permission was given to run the fire-breathing 427 SOHC, but a weight penalty would be imposed on any vehicle with the SOHC engine. In response to the decision, Henry Ford II pulled his company out of NASCAR competition on April 15. The following July, Ford Motor Company announced a limited return to NASCAR. From Mercury's standpoint, this worked out wonderfully. Iacocca had, under corporate orders, stayed out of racing in the first portion of the season. When the all-clear was sounded, he was ready with a different type of Mercury than in years past. With Dieringer behind the wheel, a 1966 Comet hit the tracks running, and it won the Southern 500 at Darlington, South Carolina, with an average speed of 114.83 miles per hour. The $12,000 race car was fitted with a 427-ci side-oiler rated at 520 horsepower with a single four-barrel carburetor and 535 horses with the dual quad setup. This racing mill, built by ace engine builder Mario Rossi, redlined at 7,000 rpm and had a life expectancy of

Mercury's King of the Drag Strip: "Dyno" Don Nicholson

Nothing in his early childhood would have hinted that he would grow up to be one of the best-known drag racers the sport has known. His family moved to California from Springfield, Missouri, where his older brother Harold gave Don a '34 Ford when Harold went into the military service. Don started hopping the car up, installing a 216-ci six-cylinder engine and other speed tricks. He even drove this vehicle flat-out on the El Mirage dry lake bed, coaxing the rod up to 118 miles per hour. He found that he was beating the local cars, so he started competing at the drag strip. Along with Harold, he opened a speed shop in Monrovia, California. When the dynamometer was sold to a local Chevrolet dealer, he used the dyno to tune his race cars and take them to the strip to do battle. He was in the Southeast when drag strip operators would see the notice on the side of his car, "Dyno-tuned by Don Nicholson." The announcers picked up on that and gave him the nickname that is recognized to this day.

He was watching a race that his friend Troy Ruttman was driving in. He met some Mercury executives track-side and was invited to Dearborn to talk with them about being a factory driver. Nicholson remembered, "I was in the office talking with Fran Hernandez and his bunch when word came down that President Kennedy had been killed." He recalled that Al Turner was like a "first sergeant," running the team day-to-day. He started racing a Comet station wagon, equipped with 427-ci engine, and was running well in it. But his teammates complained, saying he had an advantage due to the wagon's shorter wheelbase and extra weight in the rear. So in late 1964, he switched to a Caliente. Said Nicholson, "Basically, I did my own development. Mercury provided us with the equipment needed, but we did the work ourselves." By 1965, he was using the famed 427 SOHC engine. With a foue-speed manual transmission, he was running in the 9-second range in a vehicle he described as "exciting to drive."

Then Nicholson invented the type of race car that would take drag racing by storm, the "flip-tops." Running a fuel-injected *Cammer* engine on 90-95 percent nitro, he was able to cover the quarter in the 7-second range in his "Eliminator I." By 1968, he was racing a '68 Cougar Funny Car with a supercharged SOHC powerplant, resulting in elapsed times in the low 7-second range and trap speeds in the neighborhood of 190 miles per hour. In 1969, he got out of Funny Cars, citing the fact that he "got a bit warm too many times." So in 1970, he got together with Grumpy Jenkins and others to start the Pro Stock series. "I felt like I could get back in a door-slammer and still make a living and have some fun racing, so that's what I did." And the racing bug is alive and well with Nicholson, as he is getting a Pro Stock truck ready for passes down the quarter. I doubt we have seen the last of "Dyno" Don Nicholson.

Don Nicholson, legendary tuner and straight-line racer, stands with a Ford Thunderbolt at the 1994 "Fords Forever" annual bash at Knott's Berry Farm in Buena Park, California.

Gritty, Bold, and Enormously Talented: Parnelli Jones

Put four wheels and a seat in it, and this man could make it win. There are few forms of racing that Parnelli Jones did not try and did not excel at. Born in Texarkana, Arkansas, on August 12, 1933, his family moved to Torrance, California, when he was a boy. There he learned to drive, emulating the stunt drivers he saw by wrestling a car through the fields around his house. Parnelli notes that "I've always been invited to race in upscale racing. It kind of directs your career." He explained why he has been in so many types of race cars. "I've always been the sort that likes to see what's on the other side of the hill, so driving a different kind of car or different tracks is something I've always enjoyed."

He must have seen a lot of hills, because he was involved in everything from Indy cars to sprint cars, midgets, stock cars, Trans-Am cars, and Off-Road cars. Over the years he amassed an incredible resume of wins. He has six Indy car victories, 25 sprint car checkered flags, 25 midget wins, and 13 visits to Victory Lane in stock cars, crossing the finish line first in the 1964 Indianapolis 500, and winning the Pikes Peak Hill Climb in 1963 and 1964.

His association with Bill Stroppe was long and fruitful. Stock cars, dirt racing, they were an unbeatable team. From off-road racing in Mexico to hill climbs, the pair found the winner's circle more often than not. His first Pikes Peak win came on his second attempt, in a Stroppe-prepared Mercury, but he had mixed feelings about the event. "I didn't like it all that much. What happens is that in the first half of the thing you're in the trees and if you get off the road, your chances of survival are great. When you get up to the last half of the race, there are a few places if you run off, well, you might as well call up the helicopter tomorrow or the next day, 'cause it wouldn't make much difference."

Parnelli was teamed up with Dan Gurney in the 1967 Trans-Am Cougar effort that came within two points of defeating the season-winning Ford Mustangs. When the season ended, so did FoMoCo's support for the team. But as Jones puts it "in their own best interest, why would you have two brothers fighting each other?" Later, racing for Holman-Moody gave Parnelli more trips to the winner's circle. But when asked whether the Fords and Mercurys felt any different from each other, he replied, "when you get inside one of them, the dashboards all look the same."

Dan Gurney piloted a Bud Moore–prepared Trans-Am Cougar during the 1967 season. Trans-Am Cougars still pop up at vintage racing events and remind fans of Cougar's glory days in road racing.

Riverside Raceway in 1985 is the scene of two Mercury Capris fighting for position. Trans-Am racing was usually a bit more brutal than the "upper-level" race classes, providing the fans with exciting shows. *Matt Stone*

Star of the Circle Track: Cale Yarborough

This native of Sardis, South Carolina, started his racing career sweeping floors at Holman & Moody, and by the time he retired as a driver in 1988, he had 83 victory trophies from his time on NASCAR tracks. Born on March 27, 1939, as a lad he sneaked into the racetrack down the road a piece to see the first running of the Darlington 500 in 1950. From then on, he knew that he wanted to be a race car driver. He entered his first race in 1957, though he was underage. Seems he slipped behind the wheel of a friend's Pontiac when officials weren't looking, finishing 42nd.

He tried out for a slot in a Holman & Moody car in 1964 at Ashville-Weaverville, North Carolina Speedway. Even though he finished 20th, he was hired to be a "gopher" at the H & M shop in Charlotte, North Carolina, earning $1.25 an hour. In 1965, he got to drive a '65 Ford Galaxie to a second-place finish at Rockingham. Holman & Moody saw that their gopher was doing well on the track and raised his pay to $1.75 an hour. In 1966, he was able to get in the top-five finishes for quite a few races, but it wasn't until the end of the season that the Wood Brothers team hired him to drive.

Yarborough started to win big on the factory-backed team. Winning on the superspeedway at the Atlanta 500 was the start of a long succession of wins that would eventually put more than $5 million in his pocket. In 1969, he was behind the wheel of the Mercury Cyclone Spoiler II and raked up six wins in two seasons. Mercury honored him with a "Cale Yarborough" Signature Edition Cyclone Spoiler, available to the public. Mercury had a Sports Panel of drivers who went around the country to auto shows talking to the public and the press. He was given a Signature Edition Cyclone to drive for a year, but he gave it back. Yarborough recalled, "I should have kept that car. Probably worth some money today." But he did get hold of two other Cyclones just like it; they are in his garage today.

Cale Yarborough prepares to get in his 1969 Mercury Cyclone Spoiler II. Notice the extended nose sheetmetal, which helped increase the speed in the "Aero Wars." Daytona Racing Archives

about 600 miles. When the Comet was racing on the short tracks, a 405-ci engine was slipped into the engine bay. Rossi said that the 500-horsepower 405 motor was nothing more than a sleeved-down 427.

When the Ford factory teams returned to compete in 1966, they took a cue from Moore, and started running the intermediate Ford Fairlanes. Mercury did not see any more NASCAR victories that year, but little did anyone know that within two years the Big "M" would become a force to be reckoned with in NASCAR. The Mercury Man would have to wait until the 1968 season to visit Victory Circle again, but when he did, it was with a vengeance. Mercury's Cyclone production car sported a new body style for 1968, incorporating a fastback design that raised the importance of aerodynamics to a new level. The intermediate body of the Cyclone lent itself to speed in a big way.

A Legend on the Road and Track

Cyclone, Spoiler, and Spoiler II

5

Cyclone. For non-Mercury enthusiasts, it doesn't conjure up mind-bending performance, but there was a time when it was highly respected on the street. Due to sensational performance on the racetrack, the Cyclone earned a performance reputation. This top Comet was a versatile platform, and Mercury used it to pursue a young clientele. When the Comet/Cyclone boldly re-entered the intermediate market, Mercury was not intimately familiar with the class. However, no one in the marketing department could be accused of being slow learners. The nameplate was on its way to a promising future and the Lincoln-Mercury Division heavily depended on the Cyclone, and later the Cougar, for sales success. In addition, these cars proved Big "M" could build attractive, agile, and exciting cars—high-performance musclecars that were in the same league as the Chevy Camaro, Ford Mustang, and Dodge Charger (not the Buick LaSabre). With the horsepower wars in full swing, and Mercury's involvement in motorsports accelerating, the future was looking very good for "interesting" products.

The Cyclone Enters the Market by Storm

When the spectators streamed into the Indianapolis Motor Speedway in May 1966, they saw a bright red Mercury Comet Cyclone GT convertible pacing the field. For the first time, the Cyclone housed the reliable 390 V-8, replacing the 289-ci mill as the top offering.

The wheelbase measurement for the entire Comet line except for the wagon was increased 2 inches to 116 inches. New sheet metal graced the fully-unitized body with its isolated front torque boxes. It created a stiffer chassis for sharper handling. The vertical headlamp arrangement carried

The new-for-1970 nose on the Cyclone Spoiler made it difficult to judge exactly where the front end was. The optional hidden headlights were a good look on this model. This Spoiler is painted in the subtle Competition Gold.

over onto the new body, as well as the full-width rear grille treatment. The body lines were smooth and had a hint of the slim-waisted "Coke Bottle" look that General Motors incorporated in their styling. Increasing the compact to intermediate-size provided additional room for larger engines. The Cyclone was offered in two body styles: a two-door hardtop and a convertible. One step up from the base car was the Cyclone GT, which used the same two body configurations. External visual differences between the two was limited to wheel covers, tape strips, and badging. The reworked interior had a padded and hooded instrument panel gracing a new dashboard. The low-reflectance hair-cell vinyl cover minimized glare. Features to meet the new government regulations were apparent throughout the passenger compartment, which included mandated padded sun visors, four-way flashers, and an outside side-view mirror. However, it was not these external details that attracted attention to the Cyclone. The Cyclone line's real attraction was demonstrated when the accelerator pedal was firmly depressed.

The Cyclone's 390 Power Play

The change to intermediate status was attained through an increase in all dimensions, including the size of the engine bay. As a result, it was possible to offer three variations of the potent 390-ci engine with the new Cyclone. The 390 powerplant joined the 289-ci engine, that had been the top of the performance heap in 1965. When a buyer ordered a Cyclone, the standard engine was the 200-horsepower two-barrel small-block. With 9.3:1 compression and a hydraulic cam, its diet of regular fuel produced 282 foot-pounds of torque at 2,400 rpm. The 390-ci engine was not offered in the standard Cyclone. If a 390 engine was desired, one had to order the Cyclone GT. This Byzantine option lineup would only get more confusing in coming years. If the standard Cyclone 390 V-8 (not available in Cyclones) was purchased, the output of the two-barrel engine differed depending on the transmission choice. A manual transmission, for example, was hooked up to a 265-horsepower 390 engine that produced 401 foot-pounds of torque. This helped the 3,315-pound Mercury step out smartly. Alternatively, a 390 with the Multi-Drive Merc-O-Matic automatic transmission was fitted to a 275-horsepower 390 with a whopping torque increase of 4 foot-pounds. Ultimately, in the Cyclone GT package, the 390 engine kicked out stronger numbers, 335 horsepower at 4,800 rpm with a torque rating

of 427 foot-pounds, from a four-barrel carburetor and 10.5:1 compression cylinders. Originally a Ford-built carburetor (GPD) was to have been installed. But the engineers changed their minds following a drag race set up for Don Frey, Ford Division Assistant General Manager with a 1964 Pontiac GTO, a GPD-equipped Cyclone, and a Holley-equipped Cyclone. The Holley-equipped Cyclone handily beat both the GTO and the GPD Cyclone. All Cyclone GTs came standard with a heavy-duty fiberglass hood with nonfunctional twin hood scoops, while this hood was optional on the Cyclone.

The Cyclone GT garnered a lot of attention and praise by the automotive press. It was named "Performance Car of the Year" by *Super Stock & Drag Illustrated* magazine. In the April, 1966 issue of *Car Life* , a Cyclone GT was tested, equipped with the optional "SportShift" C-6 derived automatic transmission, which allowed the driver to manually change gears. When fitted with this option, the Cyclone was called a GTA. The performance numbers they compiled were in the middle of the "hot-car" pack, with a 0-60 time of 6.6 seconds, while running the quarter-mile took 15.2 seconds at 90 miles per hour. *Car and Driver* (March 1966) ran a 1967 390-equipped Cyclone GT down the quarter-mile. With its 4.11:1 rear axle ratio, it covered the quarter-mile in 13.98 seconds at 103.8 miles per hour. Still, these must have been attractive numbers to a lot of people, as 13,812 Cyclone GT Hardtops and 2,158 Convertibles cleared the showroom.

The Cyclone's Second Year

In the 1967 model year, Cyclones received minor trim and revised powerplants, but significant changes loomed down the road. To increase the buyer's confusion, the myriad of engine options had increased. The standard Cyclone came with the 200-horsepower 289 engine that produced 282 foot-pounds of torque at 2,400 revs. The two-barrel carburetor fed fuel into 9.3:1 cylinders, while a single exhaust led the gases away. The optional engine upgrade, the mildly tuned two-barrel 390 V-8 that had 9.5:1 compression ratio, produced 270 horses at 4,400 rpm and a torque rating of 403 foot-pounds at 2,600 rpm.

The Cyclone GT had a Marauder 390 GT under the fiberglass hood, and horsepower fell to 320 at 4,800 rpm from the previous year's 335 horsepower. That was the sole engine for the Cyclone GT. However, there were a couple of engines that could be shoehorned into the engine bay if you did not mind driving a non-Cyclone GT. The rest of the Comet

Stacked headlights debuted in 1965 on the Cyclone and gave the mid-size vehicle a much more Ford look. The vehicle did not look like it weighed 3,074 pounds. The grille was unique to Cyclones, differentiating it from the Comet. *Ford Motor Company*

two-door lineup, from the regular Cyclone to the bottom-feeding 202 two-door Sedan, could be fitted with one of two optional big-blocks. The 427 Cyclone kicked out 410 horsepower at 5,600 rpm, while the torque rating of 476 foot-pounds would waste a set of rear tires in a few determined moments. Super Premium fuel was mandatory with 11.1:1 compression and solid valve lifters. The four-barrel carburetor fed the gas in, while dual exhaust helped the de-tuned racing engine to exhale. Stronger yet was the Cyclone Super 427 V-8, its dual four-barrels raising the horsepower to 425 at 6,000 rpm and the torque to 480 foot-pounds at 3,700 rpm. With both monster engines, a four-speed manual transmission was the only unit in town. At the other end of the driveshaft, the rear axle gear was 3.89:1, period. How many of these beasts were actually put into the garages of buyers is not known, but probably not very many. Grocery duties were not encouraged; it was a purpose-built vehicle, that purpose being the drag strip. The fact that it had a license plate frame was a minor inconvenience.

Overall, Mercury was doing well in various racing venues, and the thrust of its model promotion was the new-for-1967 Cougar. The 427 engines were offered to satisfy the sanctioning body's request that the racing vehicles be production based with powerplants available to the public. Racing really did help "sell on Monday." And with the Ford Motor Company into "Total Performance," the public was the winner, whether they were following Mercurys on the racetrack or putting one in their garage.

The fiberglass hood was a step in the right direction from a performance standpoint. It was also cheaper to make, with the twin scoops, than a steel unit. *Super Stock* magazine was certainly impressed with the entire Comet line in 1966. *Ford Motor Company*

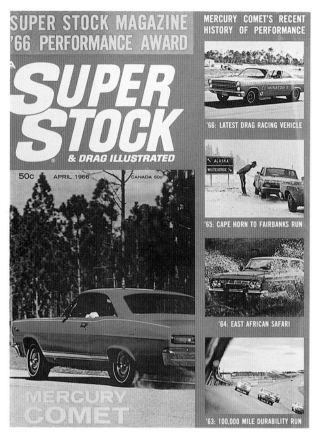

The April 1966 issue of *Super Stock* presented the history of the Comet to date. Mercury got an incredible amount of mileage out of the magazine's award, using it in a score of ads. *Petersen Publishing*

The Cyclone's New Identity

When the $2,768 1968 Mercury Cyclone debuted, it did not look anything like its predecessor. It was not considered an upscale Comet anymore. In fact, the Comet was the entry-level Montego, which looked like a full-sized Mercury that had been left in the dryer too long. The Cyclone had its own identity and was distinctly different from the Comet, but it was the top of the line Montego. Confused? Wait until we come to engines. Like the 1967s, Cyclones came as a standard model and the Cyclone GTs were the upscale version, and both still had 116-inch wheelbases. But that was where the similarities with the old body style ended. The new body looked faster, especially in the new fastback style, and on the track it proved to be faster. The body was offered in two styles, the formal hardtop and the new fastback. The fastback outsold the hardtop by a ratio of 20:1. It was no mystery. It shared the basic body-shell with the Ford Torino, and it had the Mercury touches such as better upholstery, more sound insulation, and Mercury-worthy trim. Even though it was basically a re-body of the 1967 Comet, it did not bring the 1967 engines with it.

The engine package was dependent on timing as much as buyer needs and wants. For 1968, the entry-level and $2,936 GT Cyclones came with another new-for-Cyclone engine—a 302-ci mill. This powerplant used an Autolite C8AS-9510-M two-barrel carburetor and a hydraulic camshaft to produce a docile 210 horsepower at 4,400 rpm. The vehicle weighed 3,254 pounds and spun out 300 foot-pounds of torque. One step up was another 302 ci, but this one upped the compression a half point to 10.0:1 and dropped a four-barrel carburetor on top. This unleashed 230 horsepower at 4,800 rpm. After this, the venerable 390-ci engine was found to be lurking on the option sheet. Yet another pair of engines were on tap, and the first was equipped with a two-barrel carburetor that provided 265 horsepower at 4,400 rpm with a torque rating of 390 foot-pounds at 2,600 rpm. The four-barrel version of the 390 put out 325 horsepower at 4,800 revs. Finally, the 427-ci engine that put FoMoCo in Victory Circle so many times headed the top of the option list. Okay, this version with a hydraulic cam, 10.9:1 compression, and a single four-barrel carburetor was rated at "only" 390 horsepower, and it could only be fitted with the C-6 automatic transmission. The engine was something of a bother for Mercury. The block, made of high-nickel iron, was difficult to machine, and needed to be stress-relieved. These expensive engines had cross-bolted four-bolt main bearing caps on the center three crankshaft bearings. So in the spring of 1968, the 427 engine exited Mercury's stage.

The old warrior was replaced by a smaller displacement 426.544-ci engine. But thanks to corporate wisdom, the new engine was called a 428. The 428 was a continuation of the "FE" family of big-blocks and a real melting pot of parts on the shelf. It had one easy-to-machine iron block, the old 406's bore (4.13 inches), the Mercury 410 stroke (3.98 inches), and the 406 heads fitted with bigger valves. The lifters were acquired from a 390 and a mild hydraulic cam. The sum of the parts was no disappointment. In fact, it was far better suited for street duty than the 427. While the old 427 engine supplied incredible power at high rpm in racing conditions, the 428 pumped out torque in effortless gobs. This mill came in two flavors: Cobra Jet 428 and the Super Cobra Jet 428. The Cobra Jet was rated at 335 horsepower at 5,200 rpm with 10.7:1 compression. It had two-bolt main bearings and a single Holley 735-cfm dual-feed carburetor. Its wicked thrust out of the hole—

Once again, Mercury was pacing the Indianapolis 500, this time in 1966. To commemorate the event, one of the flashiest Mercurys ever, the official 1966 Indy 500 Cyclone pace car, was sold to the public. *Ford Motor Company*

The Cyclone featured a new body for 1968 with smoother lines, and fastback hardtop was available. The NASCAR version of this production car had marginally better aerodynamics than the Ford Torino. Thus it became the stock car of choice for many. With its new fastback styling, the 1963 Marauder was the forerunner of the muscle car. Its sloped rear window was a benefit to racers, and a 427 side-oiler could be slipped under the hood. *Ford Motor Company*

445 foot-pounds at 3,400 rpm—made any street racer stand up and take notice. *Motor Trend* (August 1968) drove one off of the streets and right onto Orange County Raceway. After removing the power steering belt and changing to BF-32 spark plugs, it pulled a 13.86 quarter- mile at 101.69 miles per hour. With a vehicle weight as tested at 3,880 pounds, the 0-60 times read like a metronome: 6.1, 6.2, 6.0, 6.2, 6.2.

The second engine, the Super Cobra Jet 428, was built for assaulting the drag strip. If you checked the box for either the 3.90 or 4.30 rear axle ratio, bingo, you took home a SCJ 428. This engine came with an external oil cooler, lightened valves, a seven-quart oil pan, and higher-volume pump. Multiple carburetors were available, and many of the 427's racing parts were fitted to this monster. The actual output varied, but the factory claimed that this engine developed 340 horsepower at 4,600 rpm and 462 foot-pounds at 2,800 revs. Ram-Air induction was optional, but it didn't

produce actual horsepower gains. However, it was an interesting option. When the accelerator was buried in the plush nylon carpeting, a small flap opened on top of the air cleaner, allowing cooler, denser air directly into the carburetor ventures. When the production totals for the 1968 model year were tallied up, 6,105 GT-optioned fastback Cyclones were purchased compared to 334 notchback models. Mercury suits quick to spot a trend killed the notchback for 1969.

The 1969 Cyclone's Plethora of Engines

The 1969 Cyclone ($2,771) came in any body style you wanted as long as it was a two-door fastback. The exterior had not changed much on the outside, except for minor trim. The vast bulk of change took place under the hood. The standard Cyclone engine was the 302 ci, 9.5:1 compression, 220-horsepower mill with its two-barrel carburetor. The 351W engine

With hood pins and a scoop, the 1969 Cyclone *looked* fast, even if the standard engine was only a 302-ci mill. Fortunately, optional engines such as a 428 Cobra Jet could replace the "little" engine. Just bring money. *Ford Motor Company*

The broad Cyclone grille could front a wide range of engines, from the standard 302 cubic-inch powerplant to a 427-ci, 390-horsepower tire shredder. This intermediate car was Mercury's smallest musclecar for that particular year, and it provided capable handling for the day.

joined the Ford family lineup in 1969. This mill, built in Windsor, Canada, hence the "W" designation, was a member of the 90-degree V-8 family of engines that started with the 221 and included the 289-ci V-8s. The 351 was available in two configurations for the Cyclone. The lower-powered version had 9.5:1 compression and a two-barrel carburetor, and put out 250 horsepower at 4,600 rpm and 355 foot-pounds at 2,600 revs. It slipped between the 302 and the 390 in the engine option list. The second and higher horsepower 351 Windsor had a 10.7:1 compression and a four-barrel carburetor. Its 290 horses at 4,800 rpm and 385 pounds of torque at 3,200 rpm made it a competitive and versatile engine for the street. The

Ford kept with its tradition of freshening body styles every two years; thus the Cyclone was restyled. Weighing a tick over 3,400 pounds, this "intermediate" was not going to leap off the starting line, but it had the passing power of a locomotive. *Ford Motor Company*

'70 Mercury Cyclone GT.
Password for action with the accent on action.

Mercury Cyclone GT is the street machine that looks race-ready. With unique running lamps, concealed headlights, sporty hood scoop, hi-back buckets, remote control outside mirror, and a 351 cubic-inch V-8 engine. If this isn't enough action for you, come give our other Cyclones a whirl. One's an unusually low-priced model with an action-hungry 429 cu. in. V-8. Then there's our competition-set Cyclone Spoiler, with aerodynamic spoilers front and rear, CJ 429 ram air V-8, Hurst Shifter*, heavy-duty running gear—the works. Mercury Cyclones—the intermediates with the accent on action. See them at your Lincoln-Mercury dealer's.

long-in-tooth 390-ci engine was still offered, but its 320 horsepower was mute testimony that technology had passed it by. The 428 Cobra Jet went unchanged and so did the 428 Super Cobra Jet option. Cyclones equipped with the 428SCJ had a different flywheel and harmonic balancer than the 428CJ. The Ram-Air option was still available, but it still gave no published assistance in increasing power. But 5,363 people were interested enough to buy one.

The Cyclone CJ buyers got only one engine to choose—the 428CJ. The engine specifications remained the same for 1969, but the four-speed manual transmission joined last year's C-6 automatic. When a customer ordered a 428CJ, it included the Competition Handling Package to cope with the massive power output of the engine. This package consisted of heavy-duty springs, shock absorbers, sway bars, and a heavy-duty clutch, all of which gave the Cyclone admirable handling qualities.

In January 1969, Mercury released two signature-model Cyclones. These special Cyclones were brought out to commemorate the competition efforts of two drivers in Mercurys the year before. Each was

MERCURY. PASSWORD FOR ACTION IN THE 70'S.

MERCURY CYCLONE *Ford*

With 10,170 Cyclone GTs sold, this Mercury version of a musclecar was considered a success. The hidden headlamps were a popular styling touch that was borrowed from the Cougar. With the large engines, the biggest problem was stopping the rear tires from spinning into a smokey mess. *Ford Motor Company*

Colors such as Competition Yellow attracted the youth market—and the police. It would take heroic restraint to keep your foot out of one of the big engines. These cars were not for the introvert. *Ford Motor Company*

offered in only one geographic region, and the Mississippi River was the dividing line. People to the east of it could buy the red and white (Wood Brothers team colors) "Cale Yarborough" Cyclone Spoiler, and people toward the west could purchase the "Dan Gurney" Cyclone Spoiler in Gurney racing colors. The only engine available was the new 290-horsepower 351-ci Windsor, and the only transmission available was the column-shift automatic. The two cars shared the same tall, winged tail spoiler, but one had distinctly different sheet metal treatments. The Gurney Cyclone Spoiler wore the standard front-end sheet metal, and the Yarborough Cyclone Spoiler featured an extended nose that required lengthened hood. The aero wars were in full swing in NASCAR competition, and this beveled front-end design proved its worth on the superspeedways. But the sanctioning body required that a minimum of 500 street copies be built so that the "new" race car was, in fact, a production vehicle. In compliance, 519 rolled off the production line. Production of CJ SportsRoof fastbacks, including the Signature models, was 3,780 units.

The Revamped Cyclone for 1970

Potential buyers browsing around a Mercury showroom in 1970 might be excused for not recognizing the intermediate-size Cyclone. The vehicle in front of them was, well, different—significantly different. The Cyclone had grown 1 inch in the wheelbase, to 117 inches, and the overall vehicle length had been stretched 6.7 inches.

Two designers, William B. Shenk and Bob Williamson, had penned the 1970 Montego/Cyclone.

Shenk had developed a full-scale clay mockup, covered it in red Di-Noc, and put it in the competition pool with the other designs. When Ford boss Lee Iacocca saw it, he said "Don't change a thing on that red car!" The design became the 1970 Ford Torino, which was the Cyclone's sibling under the Ford family umbrella. When Shenk was transferred to Ford of Germany, Williamson took over and slightly "modified" the design so that it had its own unique style and identity. The body panels were different from the Ford version. The stretched wheelbase on the clay model had simply been for aesthetics, but a longer wheelbase was desired. With Iacocca's statement not to change a thing ringing in his ears, Williamson had engineering move the rear axle back 1 inch on the long leaf springs, and reshape the roofline from the previous year. Instead of a true fastback it was now a "Hardtop Coupe," and the rear window was slightly concave. While not as aerodynamic as the prior two years, it was an attractive design. The interior was also the recipient of the stylists' efforts as well. Full instrumentation was installed on the Cyclone Spoiler, but the gauges were mounted in the middle of the dash, canted toward the driver. But as in years before, powertrain choices remained plentiful.

A trio of Cyclones were on the menu—ranging from mild to wild. An official Mercury brochure helped educate the sales staff on the differences in the models as well as pointing out selling features. From the new grille treatment to the taillamp area, Mercury had the answer: "The muscle in this muscle-car is evidenced by the unique, performance-styled grille." And "The Cyclone name, which suggests the get-up-and-go that makes Cyclone a natural for the youth/performance crowd, highlights the sporty look of the rear fender lines." So for starters, the base $3,238 Cyclone.

This was a sleeper. There was a new kid in the engine room—the 429. From its bench seat to the standard 429-ci engine, this Cyclone did not exude blatant power. This was a relatively new engine, and it was not related to the "FE" family. It came from the "385" group. This engine appeared in the 1968 Thunderbird. It had thin-wall casting design, no cylinder skirting below the crankshaft centerline, two-bolt main bearing caps, and cast-aluminum pistons. With 10.5 : 1 compression, hydraulic lifters, and the 600-cfm Autolite four-barrel carburetor pouring premium fuel into the combustion chambers, this "base" engine developed 360 horsepower at 4,600 rpm, and 480 foot-pounds of torque at only 2,800 revolutions. Like the entire Cyclone range, the four-speed manual transmission was hooked up to a Hurst shifter, and an

Cale Yarborough has two of these in his garage. Equipped with a 351W engine only, they were for show, yet the 1969 Cyclone Spoiler Signature Edition was a head-turner in its Wood Brothers racing colors.

optional Select-Shift automatic transmission was available. Ford performance figures were very informative. The 3,097-pound street missile produced a 0-60 time of 7.2 seconds, and the quarter-mile went by in 14.8 seconds at 95 miles per hour.

The Cobra Jet 429, rated at 370 horses at 5,400 rpm, provided more potent straightline speed. A 720-cfm Rochester Quadrajet carb fed the 11.3:1 compression, helping to produce 450 foot-pounds of torque. Sales people emphasized the fact that this was the consumer version of the NASCAR record-setting CJ engine, and many used the information to good effect. The Ram-Air option was available with this engine, and according to the factory, a vehicle equipped with this feature had the "ability to outperform the equivalent non-ram engine by a half-length to a full car length in ten-second acceleration tests." Ford put its machine through its paces, coming up

with a quarter-mile time of 14.2 seconds and tripping the lights at 99 miles per hour.

At the top of the power list was the Super Cobra Jet 429 engine. Though it was only rated at five more horsepower than the CJ, at 375 at 5,600 rpm, the actual horsepower output was higher. It was equipped with a 780-cfm Holley carburetor, mechanical valve lifters, four-bolt main bearing caps, and 11.3:1 compression, which cranked out 450 foot-pounds of

Next page
The rear spoiler was filched from the Mustang, but it was said to actually reduce lift. Trunk access was a bit tight, but the Cyclone Spoiler was meant to haul something else. The fastback design was a very graceful style, and its wind-cheating characteristics made the Mercury Cyclone a strong competitor on the NASCAR circuits.

Looking like a shark among the minnows, this 1970 Cyclone Spoiler packs a 429CJ engine under the functional hood scoop. With 370 horsepower under the hood, the Cyclones provided scintillating street performance.

torque. The Drag Pak consisted of 3.91:1 limited-slip rear end and an oil cooler. This Super Cobra Jet option was well worth the money. It gave the car the traction to utilize the horsepower on tap and to launch it out of the hole. If a buyer was searching for maximum acceleration potential, the Super Drag Pak was a required option. The Super Drag Pak was the same as the Drag Pak except that the rear end carried the 4.30:1 ratio Detroit Loker. Engines with the manual transmission were fitted with a dual-point distributor and an overspeed governor, set at 5,800 rpm with the Cobra Jet and 6,150 with the Super Cobra Jet. Equipping a Cyclone with an automatic transmission got the buyer a single-point distributor with no rev limiter. Ford figures reflecting the Cyclone's performance showed that this was not your normal grocery-getter. The drag strip tests came up with an elapsed time of 13.8 seconds at 104 miles per hour, with a 0-60 time of 5.9 miles per hour. But if the sound of tires protesting was not enough of an attention getter, a

buyer could move up the option sheet to the Cyclone GT. Not a big seller in 1970, only 1,695 Cyclones found a home.

GT Performance Package Variety

Once again the Cyclone GT "package" vehicle had an enormous selection of engines. Mercury marketed this vehicle as "A smooth, sophisticated performer in the GT tradition." This was the only one of the three Cyclones that came with standard concealed head-lamps, and according to Mercury literature the "Cyclone GT can be highly personalized." If the standard engine didn't satisfy the need for speed, there were many engines on the option list that could propel the car to dizzying speeds. The standard engine in the 1970 GT was the 351 Windsor, which was rated at 250 horsepower at 4,600 rpm and 355 foot-pounds of torque at 2,600. With a two-barrel carburetor and 9.5:1 compression, the GT provided capable V-8 performance. But standard engine performance increased

This functional hood scoop has a special air cleaner lurking under the hood. This differentiation made the 429 Cobra Jet a 429 Super Cobra Jet.

significantly by mid-way through the model year when the 351 Cleveland replaced the 351 Windsor. The 351 Cleveland engine developed 300 horsepower at 5,400 rpm and 380 foot-pounds of torque came via a 600-cfm Autolite carburetor feeding premium fuel to the 11.0:1 cylinders. Dual exhaust allowed the engine to breathe, but the Cross-Country Ride Package suspension did nothing for "muscle-building."

The 429-ci engine was the next step up on the option list, and it was the same engine that appeared in the standard Cyclone. To handle the big-block's increase in torque, tire size grew from the F70x14s that came with the 351 engine to G70x14. In addition, the 429-powered GT received the Competition Handling Package, consisting of heavy-duty springs and shocks and a larger front stabilizer bar.

Next stop on the horsepower option train was the CJ 429, which was no different from the same engine in the Cyclone. The same 370 horsepower, the same 450 foot-pounds of torque. Ho, hum. And like the standard Cyclone, top dog in the engine bay was the Super CJ 429. Which still meant that Joe Buyer was also paying for the Drag Pak option, no weaseling out of it. With the standard bucket seats and dual racing mirrors, this was a vehicle meant to compete with luxury "musclecars" of the GTO class. Mercury said that the Cyclone GT put forth a "head-turning, image-building appearance." Evidently, quite a few people needed to work on their image, as Mercury sold 10,170 GTs in model year 1970.

Last on this illustrious list was the $3,759 Cyclone Spoiler, "The Muscle Intermediate that Goes All the Way." This was the leader of the Action Pack. And it

was not lacking for style. With a chin and adjustable tail spoiler and side striping, it was hard not to notice this Mercury, especially if painted in one of the three "Competition Colors." The interior came with high-back bucket seats with integrated head restraints. The concealed headlamps from the GT model was an option. The standard instrumentation group gave the driver full readings from under the hood, as well as a 140 miles per hour speedometer. Speaking of under the hood, the Spoiler came with only two engines. The base unit was the Cobra Jet 429, its 370 horses intact. Ram-Air induction was standard on the Cyclone Spoiler, as was a 3.50:1 rear axle ratio, Traction-Lok differential, and a four-speed manual transmission with a Hurst shifter. For those buyers who felt that this engine was insufficient for their needs, the Super CJ 429 with Ram-Air induction was the only optional

The engine compartment of a 1970 Cyclone Spoiler, equipped with a 429CJ and functional Ram Air. This engine filled the space between the grille and the firewall, and under full throttle it provided a glorious exhaust note. Notice the placement of the horn on the shock tower.

The 1972 Montego GT was the swan song of Mercury performance in the halcyon days of the early 1970's. Equipped with a 351C as standard, the GT only sold 5,820 vehicles, not enough to keep it around for 1973.

engine in the Spoiler. That meant the Drag Pak was also coming home. But only 1,631 were built, so Mercury kept an eye on sales for the following year.

The Cyclone's Last Call

Change was not sweeping through the Cyclone lineup for the 1971 model year. Yet. The GT's gunsight front grille was slightly modified, and new full-length tape stripes went down the sides. The biggest news for the Cyclone line was the shrinking engine list, and rising prices. Government-mandated emission controls were becoming a nightmare for the engineers, as the technology of the day was a bit behind the clean-air dreams of Washington. Besides, Ford Motor Company had pulled out of supporting motorsports in 1969, and with no direct involvement, its building of hyper-powerful street cars was making less and less sense. And cents.

Three Cyclones were still available, with the standard $3,369 Cyclone's base engine now the 351

Cleveland mill. Unfortunately, the engines started losing lots of power. With 10.7:1 compression and a four-barrel carburetor, its output was down 15 to 285 horsepower at 5,400 rpm. Torque stayed the same at 300 foot-pounds at 3,400 revs, but the future was looking a little dim for performance fans. Two 429 engines were still offered, the 360-horsepower version being dropped from the lineup. The Cobra Jet and Super Cobra Jet were unchanged at 370 and 375 horsepower respectively. But buyers were not impressed with what they saw and stayed away in droves. Only 444 standard Cyclones were sold in 1971, not the kind of sales that puts a smile on a bean counter. Maybe the GT model, selling for $3,680, could pull the sales numbers up. And it did, a little, with 2,287 selling. The GT came with the 351W engine again, sporting a two-barrel carburetor. Now the two-barrel carburetor–equipped engine put out only 240 horsepower, still at 4,600 rpm. Dropping the compression half a point did nothing for the torque, which fell to 350 foot-pounds at 2,600 rpm.

Sharing the body shell of the Ford Torino, the 1972 Montego GT was Mercury's version, with upgraded trim and improved appointments. The Cragar SS wheels, while not stock, are what most hot rodders of the day were installing.

The $3,801 Cyclone Spoiler was not immune to the effects of the shifting around of powerplants. It now came standard with the 351 Cleveland engine, with the CJ 429 and Super CJ 429 available as options. While the Cobra Jet was left alone, the Super CJ's compression was lowered to 11.0:1. These limited choices did not endear the Spoiler to the hearts of the public, who bought a paltry 353 units for the entire model year. Total Cyclone production in 1970 had been 13,496, while total sales in 1971 were only 3,084. It didn't take a rocket scientist to foretell what Mercury was going to do with the Cyclone. As a line of production performance automobiles, it was time to fade to black. Interest in big musclecars had waned, and the oil embargo in 1973 put the final nail in the coffin. But while they were being built, the Cyclone line gave a great accounting of themselves. They were American Muscle from the Big M.

The Cyclone GT was no longer a separate model, but now a GT Appearance Group. Bucket seats, remote-control driver-side racing mirror, and a rim-blow steering wheel were a few of the items that were included with the package.

The standard 351C engine lacked the visual impact of a 429CJ, and it also lacked the power, thanks to emission regulations. With 3,517 pounds to lug around, acceleration and top speed were lackluster to say the least.

The Luxury Sports Car

Cougar 1966-1968

6

In 1966, boxes of Cougar Crackers were shipped to media types around the country. Shaped like little cougars, they were part of the public relations campaign that C. Gayle Warnock used to introduce the new Cougar. From Cougar, Washington (population 13), promotional material was sent out, including seasoned ground beef "Cougar Burgers." Cougar Fortune Cookies from Hong Kong, with the message, "Man who have Cougar never lonely," went out to spread the word. But before the PR campaign, advertising gimmicks, and the media hype, the Cougar had a long gestation period prior to being launched. In early 1963, the "Model T-7" was started as a way to re-body the planned Mustang floor pan. As a smaller Mercury, it was intended "for the man on his way to a Thunderbird." Ralph Peters, in charge of the Cougar Planning staff, decided that the Cat would have proprietary sheet metal rather than just changing the fenders and end caps. High performance was not part of the initial vision, but eventually it would be part of the game plan.

When the first Mustang was released, the Cougar project was born. On Friday, April 17, 1964, Hal Sperlich's and Lee Iacocca's four-seater Mustang hit the showroom floors, creating a sensation around the country. Before Ford dealers shut their doors on that day, some 21,000 were sold. One year later, 418,812 Mustangs had been delivered into customers' hands. These numbers brought tears of joy to Ford executives' eyes. However, Lincoln-Mercury dealers were left out in the cold because they didn't have a similar pony car in their lineups. To say they were clamoring for a comparable showroom draw is an understatement. Mercury's product planners

The Cougar GT-E was Mercury's big-block challenger in the 1968 muscle car wars. It was equipped with Ford's venerable 390-horsepower 427-ci side-oiler engine. The GT-E was capable of low 6-second 0-to-60 times and quarter-mile runs in the low 15-second range.

wanted a piece of the action, but they weren't quite sure what car they wanted to create—a sports car, a luxury car, or a family car. Simply re-badging the Mustang wasn't the route Mercury management wanted to take. Countless meetings were held to determine the specific image the vehicle would present. Finally, on February 18, 1965, Ford management approved building a Mercury high-performance sports car—the Cougar, an elegant, well-dressed sports car that personified the Mercury ideal. It had European elegance with American muscle. A headline on an early Cougar advertisement read "Mercury believes a man shouldn't have to buy $800 worth of ocean to get the European look. Meet Cougar." It fit between the Mustang and Thunderbird in the marketing hierarchy, leaning toward the opulence of the Continental. With the tag line "The Fine Car Touch Inspired by the Continental," there was little doubt that Mercury was targeting an upscale clientele. The Cougar name was originally on the styling concept car from David Ash. That concept automobile evolved into the production Mustang. The Cougar name was lifted from initial design exercise and put into "storage," where it stayed until being used on the Mercury offering.

A Style All Its Own

The Cougar was softer riding, quieter, and more luxurious than the Mustang. The luxurious Cougar came out in the latter half of 1966 with the entire Mercury lineup. With its retracting headlights, sequential turn signals in triple taillights borrowed from the Thunderbird, and electric razor grille, it had real presence on the road. The interior of this first-year offering was comfortable for the day. Vinyl bucket seats, deep loop carpeting, a three-spoke Sport steering wheel, and 123 pounds of sound-deadening material gave the buyer a feel for upscale Grand Touring, American style. Cougar Design team leader John Aiken recalled that the Jaguar XK-E had a heavy influence on the interior appointments. From the wood-grain dash to the circular dials and the toggle dash switches, the British approach to upscale had been implemented on the Cougar. Or as the sales literature was quick to point out, "the Man's Car." The October 1966 issue of *Car Life* wrote, "There's a gracefulness and shapeliness about the Cougar that its progenitor doesn't have." Unlike the Mustang, the Cougar was offered in only one body style, a two-door notchback. Yet the proportions were right, enough luxury equipment was

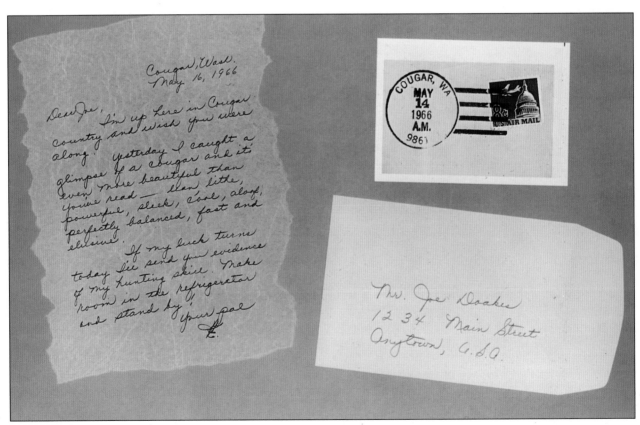

Mercury public relations man C. Gayle Warnock had this letter sent to members of the media prior to the Cougar's debut. Notice the "Cougar, WA" postmark. *C. Gayle Warnock Collection*

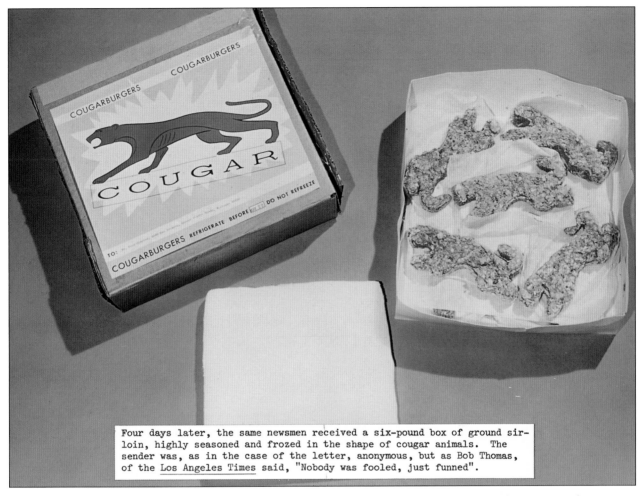

Four days later, the same newsmen received a six-pound box of ground sirloin, highly seasoned and frozen in the shape of cougar animals. The sender was, as in the case of the letter, anonymous, but as Bob Thomas, of the Los Angeles Times said, "Nobody was fooled, just funned".

Cougarburgers shipped in dry ice to media around the country for the introduction of the 1967 Mercury Cougar. The seasoned ground beef was shaped like the Cougar emblem. *C. Gayle Warnock Collection*

offered as standard, and at a base price of $2,851, value for the money was evident.

The 1967 Mustang chassis needed some modifications so that the handsome Cougar body could be fitted to the chassis. The wheelbase had been stretched 3 inches to 111. The rear leaf springs were lengthened 6 inches, and the suspension bushings were softer for a plush, luxury-car ride. To give the desired level of road isolation, the suspension was, as Bob Negstad put it, "wrapped in rubber." Where the Mustang specifications might call for a bushing of a specific durometer hardness, the Cougar part equivalent would be a handful of points softer, though almost identical. There weren't any Cougar-unique components that had not seen duty on the Mustang: this commonality helped hold costs down and speeded up the engineering. A set of 7.35x14 tires provided contact with the road, and optional front disc brakes provided a stronger, fade-resistant "sporty" feel to the brake pedal.

Unlike the Mustang, no six-cylinder engine was offered. The base powerplant was the ubiquitous 289 V-8 with 9.3:1 compression churning out 200 horsepower at 4,400 rpm with a two-barrel carburetor and single exhaust. The engine's compact dimensions were ideal for the engine box. The tall shock absorber towers did not interfere with this lightweight mill. The "GT" option provided a "Super 289" engine with a four-barrel carburetor, dual exhaust, and 10.0:1 compression. This choice put out 225 horsepower, delivered at 4,800 rpm, while 305 foot-pounds of torque at 3,200 rpm pushed the ground away from the tire tread. The GTs were fitted with a sportier than stock suspension with solid rear bushings, stiffer springs, a .84 inch anti-roll bar, and firmer springs. Three-speed and four-speed manuals, and the Merc-O-Matic three-speed automatic were the available transmissions.

For the power-hungry, the option to order was the GT Equipment Package, which automatically came

The new 1967 Mercury Cougar rode a Ford Mustang platform. However, the wheelbase was stretched 3 inches to 111 in the Mercury application. The standard engine was a two-barrel 289-ci engine, developing 200 horsepower at 4,400 rpm. *Ford Motor Company*

The factory cutaway drawing of the 1967 Cougar displays all the car's major componentry. The standard equipment includes front disc brakes, front wishbone suspension, heavy-duty shocks, and a solid-axle rear suspension. *Ford Motor Company*

This 1967 Cougar was photographed by Boulevard Photographic for official Mercury literature. Mercury was obviously targeting the older, upscale clientele. *Ford Motor Company*

with the 390 big-block engine. The original 1964-1966 Mustang chassis had insufficient room under the hood for larger engines, but that shortcoming was corrected in 1967. The option was called the Marauder 390 GT in the entire Mercury line, and it produced out 335 horsepower at 4,800 rpm with 427 foot-pounds of torque at 3,200 rpm. This engine had been developed for large passenger car use, so in the lighter Cougar, it had scintillating acceleration even though it put an additional 250 pounds over the front wheels. The engine had a 10.5:1 compression ratio, the Holley C70F four-barrel carburetor, and

The GT-E was the ultimate Cougar in 1968. With the 427-ci engine shoehorned into the engine bay, the 425-horsepower powerplant was more than a match for most of the competition. The argent-colored lower body paint was the quickest way to identify a GT-E. The hood scoop was a nonfunctional styling piece. *Ford Motor Company*

spent gases exited through a low-restriction exhaust system. A 3.00 axle resided at the rear to handle the copious torque of the 390 engine. The optional "Power Transfer" unit housed a 3.25 set of gears. The brakes were upgraded to power front discs, and 8.95x14 Wide-Oval tires tried to keep the car hooked up to the ground.

The Cougar Performane Package XR-7

In 1967, the battle for the buying public's attention heated up with rival offerings by Chevrolet (Camaro), Pontiac (Firebird), and Chrysler's restyled Barracuda. The XR-7 was introduced as a midyear model in January 1967 to go head-to-head with the mid-sized musclecar competition. From the outside, only a small badge on the roof's quarter panel indicated that this was the high-level Cougar model. Inside, there was no mistaking the luxury overtones. A fancier interior, replete with wood-grained dashboard appliqué, toggle switches, and different gauges—these were ready signs that the buyer's $3,081 was money well spent.

By the end of the model year, 27,000 XR-7s had rolled off of the showroom floor, and *Motor Trend* magazine awarded the Cougar its prestigious "Car of the Year" honor. A July 1967 *Car and Driver* road test put the XR-7 against the Jaguar 420. The 390-ci, 320-horsepower, automatic transmission example went from 0 to 60 in 6.5 seconds, compared to the Jag's 10.2 seconds. The American Cat had a 1 mile per hour lead in top speed, at 123 miles per hour. The

result: the Cougar was top Cat. The acceleration difference was enormous, and it cost 45 percent less than the Jag.

The competition potential of the new Cat was not lost on Ford. With a common belly pan stamping, parts that worked on the Mustang would likely have the same effect on the Cougars. In 1966, Ford Mustangs earned the coveted manufacturer's championship over the Chrysler Plymouth team in the inaugural year of Trans-Am racing. Naturally, Ford wanted to increase its competitiveness even more in order to win big in 1967. The new Cougars were pressed into duty with Bill Stroppe and Walter "Bud" Moore running the Mercury show. Moore hired Dan Gurney, Parnelli Jones, and Ed Leslie to drive a small fleet of red and silver Cougars on the Sports Car Club of America (SCCA) circuit. This team earned five poles and four wins, but Ford's Shelby American Mustang team were the clear favorites to win the Trans-Am title at year's end. When the dust settled, the Shelby Mustangs had secured another championship by a margin of only two points over the factory Mercury teams of Stroppe and Moore. This did not sit well with Ford, which pulled the funding plug on the Cougar team for 1968. To commemorate Dan Gurney's ties with the Cougar, Mercury released a trim package cleverly called the Dan Gurney Special. It was available with any engine offered, and it consisted of turbine wheel covers, chromed engine dress-up kit, F70x14 wide-oval nylon cord whitewall tires, and a couple of Dan Gurney signature decals.

The Cougar Stretches Its Legs

The Cougar made some crucial performance strides in its second year. There was not a lot of external differences between 1967 and 1968. The government-mandated side marker lights are the most obvious, but the real efforts were directed under the skin. Mercury borrowed powerplants, so engine selection and size grew from the Mustang engineers. Rudimentary emission controls were arriving on the scene, and the Detroit horsepower wars were swinging into full

This was the view most people saw if the driver of a GT-E pushed the accelerator pedal into the carpet. The GT-E was available with Select-Shift Merc-O-Matic transmission only. The auto tranny performed admirably, but a four-speed manual transmission option would've been desirable. The blacked-out vertical taillight ribs are unique to this very-high-performance model.

Special badging behind the front wheels signified that the area under the simulated hood scoop was full of Ford Motor Company's strongest engine in 1968. The 427-ci mill was only fitted to 338 Cougars and was rated at 390 horsepower but could be increased using dealer supplied parts.

The plush XR-7 interior was standard on the Cougar GT-E. The three-speed Merc-O-Matic automatic transmission was the only one offered with the 427-ci engine. The Jaguar-inspired dash design included toggle switches and a full complement of gauges, however, they were spread across the faux wood panel in a seemingly random fashion.

battle mode. More was better, and this concept was alive and well under the Cougar's long hood.

The 289-ci engine had grown into a 302-ci V-8. This 210-horsepower, two-barrel carburetor–equipped mill was now the entry level engine for the base model Cougar, the XR-7, and Decor Group options. For buyers wanting to get a hotter small-block, the optional four-barrel/dual-exhaust 302 put out 230 horsepower at 4,800 rpm, and a few Cougars were equipped with 289s putting out 195 horsepower. These two-barrel versions were not as popular as the four-barrel engines, and increased power (and sales) was the goal.

FE big-block engines (390 and 427 V-8) offered an easy way to increase horsepower and torque. The Marauder 390 P was the low-end offering that had 280 horsepower at 4,400 rpm via a two-barrel carburetor. Bolting on a four-barrel carb turned the engine into a 390 GT, and it put out 325 horsepower at 4,800 rpm. This was the standard engine fitted to Cougars with the GT Equipment Group. This package included stiffer front and rear springs, heavy-duty front and rear shocks, and a larger-diameter stabilizer bar for "cat-sure" handling. Front power disc brakes were also standard on the Cougar GT.

The long air cleaner was part of the package that dealers could install on the 427-ci engine, replacing the single factory four-barrel carburetor with a dual-carb setup. The aluminum air cleaner top was recessed to accept the "427" emblem, but the customer had to pay extra for the emblem itself. This GT-E is equipped with finned aluminum valve covers, about 20 sets are known to exist. These valve covers were made by Shelby and have a Shelby part number inside. *David Newhardt*

For buyers willing to fork over an extra $908, they could own a GT-E. This option was discontinued in December of 1967 after only 338 of the 427-ci side-oiler ground-pounders were sold. The 427 had been slotted under the Cougar's hood, but concessions had to be made. Equipped with hydraulic lifters, the V-8 was rated at 390 horsepower at 5,600 rpm. A mild camshaft and 10.9:1 compression did not allow the engine to realize its potential. The Low Riser version, using a low-rise-type intake manifold under a 650 cfm Holley, used heads with larger valves for improved breathing. In the theme of sporting elegance, and emission controls, the only transmission bolted to the engine was the three-speed Select-Shift Merc-O-Matic automatic. The lower body areas were painted argent, and two chrome strips the width of the blacked-out grille and rear taillight blades spoke of something out of the ordinary. On the front fenders below the Cougar emblem were handsome 7 liter GT-E emblems. A glance into the Mercury could reveal the optional XR-7 interior. The hood was graced with a nonfunctioning hood scoop.

Cobra Jet–Powered Cougars

On April 1, 1968, the 428 Cobra Jet engine was released. The 428CJ became the only engine available in the GT-E, but it could be ordered for any Cougar. This mill was a 428 that featured 427 Low Riser heads and a 735 cfm Holley carburetor fitted to a cast-iron version of the 428 Police Interceptor intake manifold. The engine produced a compression ratio of 10.6:1. The 428CJ could be mated to a four-speed manual or a C-6 automatic. This engine was conservatively rated 335 horsepower at 5,200 rpm. And atop the engine compartment was a hood with a functional Ram-Air

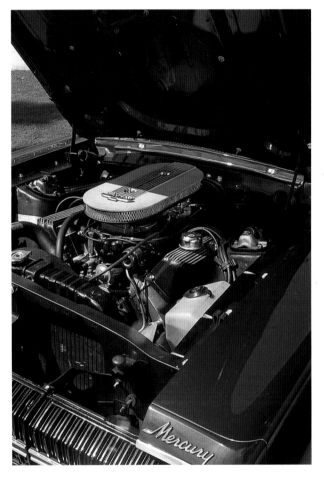

The 427-ci engine came from the factory with a single four-barrel carburetor, but dealers would often install dual four-barrel carbs per customer wishes. The massive engine in the small engine bay meant that room to work on the engine was scarce.

The 390-ci, 6.5 litre "FE" block engine was a tight fit in the 1968 Cougar XR-7's engine compartment. The mill produced 325 horsepower at 4,800 rpms and was equipped with a four-barrel carburetor. Factory air conditioning kept the occupants cool and filled what little space was left over under the hood. *David Newhardt*

feature. Interestingly, this package was not heavily promoted by Mercury because it was contrary to the corporate stance on competition. It was the company's mission to produce elegant sports cars, not mind-blowing race cars. Remember, factory-backed racing Cougars were a thing of the past. But the past had not been forgotten, the XR-7G ("G" for Dan Gurney) was released as a midyear option in the spring of 1968. It was something of a continuation of 1967's Dan Gurney Specials, but it had a bit more that visually separated it from other Cougars. A non-functional fiberglass hood scoop, road lamps, hood pins, styled steel wheels, and quad exhaust tips were exclusive to the option. It was fitted with either the 390 or 427 V-8. This was a good indication of the performance path that Mercury had chosen. But this

The hood scoop on the Cougar XR-7G was for looks only, while the race car–inspired hood latch pins helped convey a performance air. But the 390-ci engine could get the Cat under way in a brisk fashion. However, with the additional 250 pounds in the front, handling suffered.

The 1968 Cougar changed very little from the previous year. Due to government regulations, side marker lights were added, and that was the main distinguishing characteristic. The official Mercury sales catalog termed it as "the better idea in luxury sports cars." This particular XR-7 was equipped with an "FE" big-block 390-ci engine.

was not the last time that Dan Gurney's name would grace a Mercury.

Mercury's ad campaign reflected the upscale life that the "average" Cougar owner might lead, and the type of audience the company was targeting. Having a real cougar scratching the paint on the hood might not have been everyone's ideal, but it generated attention. The big cat, named Chauncey, was rather easy to work with as long as he was supplied with raw chicken necks. And the tone taken would raise the hackles of the "politically correct" today. "If you want a sports car and she wants a luxury car, or vice versa, you both want a Cougar." References to the 427-ci engine's competition history were used, and Indianapolis as well as the 24-Hours of LeMans were mentioned in the same sentence in the GT-E literature. Yet the overriding theme was luxury. Mercury literature stated a "full vinyl trim so richly supple you'd swear it was leather" and "the tawny tones of simulated walnut." This is how Mercury was able to underprice Jaguar by 45 percent, but in the tradition of Detroit, 1969 saw growth, in all dimensions.

The 6.5-liter badges on the fenders behind the front wheels and the argent-colored lower body section was indicative of a Cougar XR-7G. The 390-ci engine was equipped with a four-barrel carburetor and produced 325 horsepower. The large "FE" big-block engine was a tight fit in the Cougar engine bay.

Mercury's Musclecar Glory Days
Cougar 1969-1971

7

The Cougar followed the Mustang in drivetrain, in basic chassis configuration, and in styling. As the upscale clone of the Mustang, what worked in the pony car was usually attempted in the Mercury offering. The two cars rolled down the production line together. In the engineering labs in Dearborn, the Cougar "mules" were developed literally next to their Mustang cousins. And in 1969, the physical dimensions, horsepower, options, and price grew. "Musclecars" were at their height, and bragging rights in the quarter-mile were the yardstick. Competition for customers in the closely fought enthusiast market was red hot, as virtually every manufacturer offered outrageous big-block and small-block performance. Mercury's advertisements were blatantly oriented to high performance, as in the copy telling the reader of "Lincoln-Mercury's ZIP code: CJ 428." Some truly memorable road cars saw daylight, the likes of which we will probably never see again. Pity…

A New Profile for the Luxury Musclecar

The chassis of an automobile "shapes" many vehicles' characteristics. When the 1969 Cougar debuted, it showed that the "more is better" mentality of the bean counters and suits had found a home. The Cougar had grown 3.5 inches in length and almost 3 inches in width. The wheelbase stayed the same at 111 inches, but about 300 pounds were added. The XR-7 convertible's shipping weight was 3,539 pounds. Increased weight was never a good thing for high performance or musclecars, but it wasn't as much of an issue for a luxury car. Yet the Cougar certainly looked the sports part. And as we will see, real motive power could be had: just bring money.

The Cat's traditional styling cues were retained, such as the hidden headlights and rear sequential

The sweeping body line was similar to the Buick Skylark of the same era, but there is no mistaking a Cougar Eliminator with its unique front fascia, graphics, and spoiler.

The Cougar XR-7, like this 1969 example, used bright vertical ribs in the rear grille to differentiate it from the Eliminator option. Backup lights located on the tips of the fenders were used for ease of parking.

The 1969 Cougar XR-7 interior was plush leather, nicely appointed with a full set of gauges, and included plenty of ersatz wood. The second generation Cougar, like other musclecars, grew longer and wider than its predecessors. Power was provided by a base 302, or an optional 351, 390, or 428. The GT and GT-E were option packages rather than individual models.

taillights operated by a transistorized timing device. The proportions were pure musclecar: long hood and short trunk. It possessed a lithe, handsome quality that spoke of tasteful opulence, yet "still greater value." In the October 1968 issue of *Sports Car Graphic*, it was noted that "for '69, Cougar appears to be the most improved of all the pony Cars." A Buick-style sweep spear character line went down the side of the car, and the vehicle height lowered 1/2 inch, which gave the Cougar a broader shouldered look for the 1969 model year. Ventless door glass was introduced.

The XR-7 option primarily lived on as a collection of convenience features. In 1969, a convertible was offered for the first time and 4,024 Merc drivers took one home. GT package was not carried over for 1969. On May 8, 1969, Mercury unveiled the Eliminator, a hot limited-edition performance special. It was named after "Dyno" Don Nicholson's A/FX drag cars from the mid-1960s, but the cars were intended for straightening out the twists in the road more than for use at the drag strip. Famed stylist Larry Shinoda created the Eliminator exterior using the tail spoiler from the Mustang and producing a handsome and well-integrated design. The Eliminator option package was not available on convertibles, and the 351 four-barrel carburetor–equipped engine was base Eliminator fare. But any of the higher output powerplants could be slipped under the Eliminator's long double-panel hood.

The interior was restyled along Mustang lines and the dash padding became more pronounced. Of course, woodgrain appliqué was still available, and what's a luxury car without that? High-back seats, "Comfortweave" vinyl upholstery, and a power-operated sunroof were all available to enhance the upscale feel. Even the optional leather seats on the XR-7 contributed to "The Fine Car Touch Inspired By The Continental." The seat's increased front shoulder and hip room provided a large measure of comfort to drivers of various heights and weights. Automatic transmissions were fitted to 89 percent of the XR-7s , and 92 percent had power steering. These numbers reflected the focus of Mercury's sales efforts and makeup of the audience that bought them.

Powerplants for the Cat

The horsepower wars were in full swing by 1969. Every American manufacturer tried to one-up the competition with cubic inches, compression ratios, and carburetor size. Everyone wanted to have the fastest car. Cougars were not immune to this campaign, and a list of underhood choices were plentiful. From the two-barreled 351-ci entry level to a pair of Boss 429-equipped Cats, there was more roar or performance than any year previous.

The base 302-ci engine for 1968 was dropped from the lineup in favor of the 351W powerplant. Known as the Windsor, after the Windsor, Canada, engine plant where it was built, it was basically an enlarged 302. When it was fitted with the Autolite C9ZF-9510-A two-barrel carburetor, 250 horsepower at 4,600 rpm pushed the 3,420-pound coupe

The styled steel wheels were a popular option on the 1970 Cougar Eliminator. Measuring 14x7, they were fitted with the latest in tire technology such as Poly-Glas bias-belted rubber, which delivered a comfortable ride, especially when cold.

to less than thrilling levels of acceleration. Dropping a 470 cfm four-barrel on top of the engine, along with a jump in compression from 9.5:1 to 10.7:1, resulted in an increase to 290 horsepower. This option had dual exhaust as well to aid in breathing and aural excitement. But in the Cougar it was decidedly understressed in 1969. When the three-speed Select-Shift automatic transmission was bolted on, it gave languid but reliable service. Mercury realized this and pulled it from the list in 1970. Its departure was not missed.

Mercury's Top Big-Block Weapon

The 428 Cobra Jet resided at the top of the engine heap in 1969. Power output, durability, and the ability to face increasing emissions regulations meant that this engine would find its way into any engine box it would fit into. There were three versions of this FE engine in the option list. The first was a carryover from the model year 1968 GT-E option, when it was introduced as a midyear option. The venerable Marauder 390 "FE" engine could be slipped under the hood, but it only gave 320 horsepower, not a major leap from the top 351 engine. And the weight of the 390 degraded handling. This engine added 250 pounds over the front wheels, but its output tended to neutralize the additional weight. Conservatively rated at 335 horsepower at 5,200 rpm, its 440 pounds of torque could turn the F70x14 bias-ply tires of the day into smoky lines. In reality, this engine was putting out almost 400 horsepower.

The next step up the option list was the Ram-Air version of the 428CJ—the 428CJ-R, but this option did not officially boost the power rating. However, mashing the gas pedal to the floor opened a vacuum diaphragm–controlled trap door and cool air flowed directly into the carburetor. An increase of dense air meant that there were more air molecules to carry the fuel charge, and that raised output. Mercury denied that output was affected, but timing slips said otherwise. Hood Lock Pins were optional, and only available with Ram Air Induction. All Cobra Jet engines were equipped with cast crankshafts, and high-performance connecting rods were swiped from the Police Interceptor engine package. With a compression ratio of 10.6:1, a 90-pound cast-iron intake manifold. and a 735 cfm Holley carburetor, this engine burned large quantities of premium fuel. Like the "entry-level" Cobra Jet, the 428CJ-R engine used the hydraulic 390 GT camshaft. The stock dual exhaust system used cast-iron header style manifolds, helping to put ozone-depleting gases into the atmosphere. Either the four-speed Top Loader or a beefed up C-6 autobox were available.

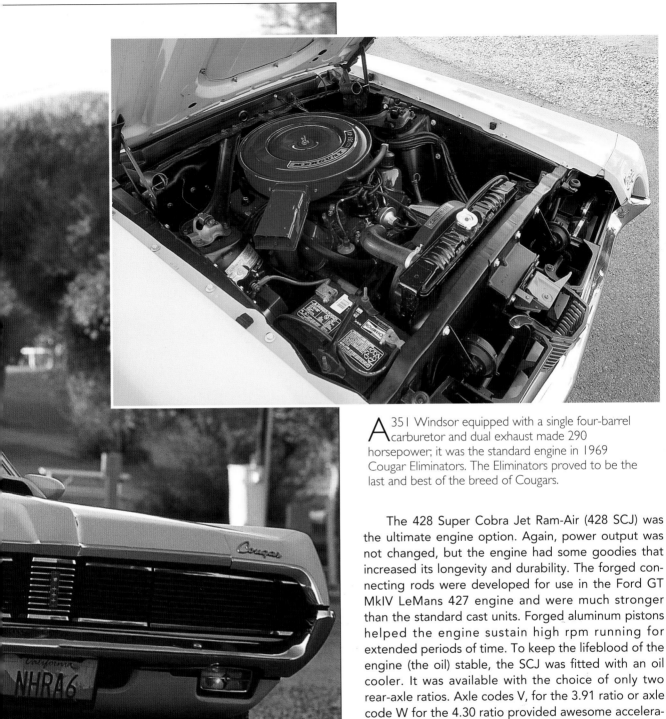

A 351 Windsor equipped with a single four-barrel carburetor and dual exhaust made 290 horsepower; it was the standard engine in 1969 Cougar Eliminators. The Eliminators proved to be the last and best of the breed of Cougars.

The 428 Super Cobra Jet Ram-Air (428 SCJ) was the ultimate engine option. Again, power output was not changed, but the engine had some goodies that increased its longevity and durability. The forged connecting rods were developed for use in the Ford GT MkIV LeMans 427 engine and were much stronger than the standard cast units. Forged aluminum pistons helped the engine sustain high rpm running for extended periods of time. To keep the lifeblood of the engine (the oil) stable, the SCJ was fitted with an oil cooler. It was available with the choice of only two rear-axle ratios. Axle codes V, for the 3.91 ratio or axle code W for the 4.30 ratio provided awesome acceleration. The 31-spline axles attached to the vaunted "N" case differential provided good rear-end durability.

The body pieces and graphics on the 1969 Cougar Eliminator were designed by Larry Shinoda's group in Ford Styling. Like the other cars Shinoda styled, such as the Boss 302, the Eliminator contains a timeless musclecar style.

All 428CJ Cougars were fitted with the Competition Handling package. This consisted of heavy-duty springs, shocks, and stabilizer bars, and if the four-speed transmission was ordered, staggered rear shocks were installed to help control axle hop.

The Small-Block Sensation Arrives—Boss 302

The Mercury marketing types required that another engine was fitted to the Cougar, courtesy of the Mustang. The Boss 302. This legendary power-plant made its mark in the Trans-Am Mustangs. This engine was a descendent of the Tunnel-Port 302. This historic small-block was offered in 1969 and 1970 for the Cougar Eliminators and the street Boss 302 Mustangs. Essentially, this was a de-tuned race engine, and it was only offered for street use due to the requirements of Trans-Am to make a set number for public consumption. With the Z-28 Camaro on the racing scene, Ford needed to counter the Bowtie's efforts. The Boss 302 was very successful in this regard. Introduced on April 1, 1969, as a midyear offering, it was Mercury's effort to capitalize on the track record of this potent engine. Rated at a laughably conservative 290 horsepower, a well-tuned Boss 302 street engine put out about 350 horsepower. To experience the thrilling rush of maximum horsepower, the engine had to be revved to 5,800 rpm. From its cross-drilled forged steel crankshaft, nodular iron fly-wheel, forged connecting rods, solid lifters, and the huge 780 cfm Holley carburetor atop an aluminum high-rise intake manifold, this engine set a benchmark for small-block performance. Fitted with Cleveland-type heads, large canted intake (2.23 inches) and exhaust (1.72 inches) valves, angled to facilitate flow and improve cooling, this powerplant had a 10.5:1 compression ratio. Forged alloy aluminum pistons with slipper-type three-ring design tended to crack within 30,000 miles.

Factory promotional literature suggested dealers point out Boss' "mechanical valve lifters for high-performance efficiency" and the "aluminum intake manifold with high flow-rate." That's corporate-ese for racing. Its factory horsepower rating was taken at 5,800 rpm, but, typically, the engines were not put under this enormous stress. A factory rev-limiter kept the engine below 6,150 rpm, although the engine was able to spin easily at 7,000 rpm. This mill could thrust

The aggressive stance was balanced by Mercury's attention to improved trim quality and distinctive styling. Graceful bumpers were not the best protection for body panels, but they certainly were attractive.

Instrumentation on the 1970 XR-7 was superb for its day. It featured speedometer, tachometer, gas gauge and passenger side clock. It addition, it had rim-blow steering wheel and Burl walnut appliqué. Note the lack of a redline on the tachometer.

a Cougar Eliminator from 0 to 60 in less than 7 seconds. The driver just needed to keep the revolutions up. One and only one transmission was available in the Boss 302-powered Eliminator—the four-speed manual. This gearbox could be ordered with a choice of close or wide ratio gear sets. And the engine could only be installed in the optional Eliminator package.

The Big, Baddest Ford Big-Block—Boss 429

At the top of the heap, another legendary engine resided, the Boss 429. The Boss 429 engine, also known as the Shotgun 429 Hemi, was installed in only two Cougars. Former General Motors executive Bunkie Knudsen was a fan of motorsports. When he was hired by Ford, he wasted no time in efforts to put together a winning race program. Originally, the Boss 429 was developed to allow Ford to race in NASCAR stock car racing against Chevrolet's COPC Mk IV 427-ci, 425-horsepower mill and Chrysler's 425-horse-power 426 Hemi. A couple of engines found their way into Cougars. Kar Kraft, of Brighton, Michigan, was the "Skunk Works" for Ford. The company performed high-end engineering and machine development for special applications—racing. This firm would stuff, shoehorn, and cajole frighteningly powerful engines into Ford products. In addition, they would rework other major components and systems, such as suspension and cooling. Kar Kraft built brutal machines to order, and produced the limited-production 1969 and 1970 Mustang Boss 429s.

The hood scoop was functional on XR-7s harboring a 428CJ with the Ram Air option. But unlike the Eliminator, there was no black tape appliqué. The "electric shaver" grille was chromed in the base and XR-7 models.

The two Cougars that received the Boss 429 engines are, without a doubt, the rarest and most desirable high-performance Mercury's ever built for the street. These engines were not related to the 427/428 FE big-block. The 385 family of engines w as destined for use in Ford's large passenger cars. When a set of aluminum "quench-Hemi" heads was fitted to the blocks, a brutal amount of power was produced. The thin-wall cast-iron, short-skirted block held a forged crankshaft using four main bearing caps. This configuration was not like the Cobra Jet which used four-bolt main caps on the center three journals only. The first 279 Boss 429 engines were built with hydraulic lifters and cam, but mid-1969 Kar Kraft started using mechanical lifters and cams. Using copper "Cooper Rings" around the cylinders and O-rings at the water passages eliminated the conventional head gasket. Resting on top of the heads was an aluminum dual-plane intake manifold with a 735 cfm Holley. With a high 11.3:1 compression ratio, only high octane fuel could be used. According to the official Ford spec sheet, the engine cranked out 375 horsepower at 5,200 rpm with 440 pounds of torque at 3,400 rpm. However, the engine hit its peak at about 6,000 rpm, and the horsepower and torque ratings were, shall we say, conservative.

The Cougar's Finest Hour

Many consider 1970 to be the finest year in the Cougar's history. It was a year of refinement, and Mercury marketed the entire package to a younger clientele than it had ever done in the past. Unbeknownst to John Q. Public at the time, it was the last year of true "high performance." Mercury had maintained a winning profile in various types of competition. It was time to see that "Winning on Sunday" translated into "Sell on Monday." Sales of Cougars had been declining since its inaugural year, and only 100,069 units sold in 1969.

A bulletin Mercury sent to its dealers said this about the high-performance market: "For 1970, more muscle is what it's all about. Engine sizes and power ratings are up, and the buyer has more models to choose from. The high-performance market is entering the new decade with potential for sales and profits far beyond the remarkable success story of the 1960s. Performance is where it's at—for buyers of all ages and backgrounds. And, in 1970, Mercury is where the action is. The Action Pack puts it there."

Mercury wrote this in dealer literature about the Eliminator: "A car completely different from the Cyclone intermediates, Cougar Eliminator fills a very important position in the Action Pack. Its size gives it considerably more visual sports car flavor than the Cyclone models. This size difference is also noticeable with the Boss 302 Engine. Ordered with this engine, Cougar Eliminator gains the excitement most often associated with European cars of the GT (Grand Touring) class." In many ways, the Eliminator comes off as a personal car. Distinctive appearance is a very important selling point. Even the name emphasizes the point.

Mercury well recognized that customers coming into its showrooms knew high performance. The literature targeted toward its sales force continued: "The performance buyer is extremely interested in his kind of car, knows a lot about it, and likes to talk about it. Share that interest. When he walks in the door, chances are excellent he already knows exactly what he wants. If he has questions, they'll be very specific. As for the question of taking used musclecars in trade, a fair allowance on such a car is a good investment for the dealership, not only for the future profits from retailing it, but also for the traffic that a good display of used musclecars will bring."

The plush XR-7 was not often thought of as a musclecar, but one look in the engine compartment would change minds. Rated at 335 horsepower, the 428 Cobra Jet with 4.13 x 3.98 bore and stroke became a legend on the street.

Only 55 1970 XR-7 convertibles were built with the 428CJ engine. This fine example of the rare Cougar has a center grille bulge, and it is 2.3 inches longer than the 1969 Cougar. The Cougar received high-back bucket seats for 1970.

A truly beautiful automobile, the 1970 XR-7 has a low stack height, which means that rear vision is free of obstructions. Argent-painted wheels were a handsome factory offering.

The same publication educated a sales force used to pitching large family sedans: "What is a Muscle Car? To begin with, it should be thought of as a specialized vehicle with far better total performance than the typical family car. In this context, 'performance' is not limited to merely power—it's much more. The muscle car is also designed and built to excel on acceleration, handling and braking. It's a 'fun' car. The typical performance buyer likes cars and he likes driving—the basic transportation aspect doesn't particularly interest him. It follows from this that the muscle car does not rely on factors such as economy, comfort and luxury for its appeal."

Sales people were encouraged, that "to increase your background, it is strongly recommended that you thumb through the enthusiast magazines such as *Car Life, Motor Trend, Car and Driver,* and *Hot Rod* on a regular basis, and back up your reading with an occasional visit to a drag strip." Still true words today. It's interesting to note that in the same publication, the Cougar is "Still the best equipped personal car in the Sports Specialty Market." So is it any surprise that the sales people as well as the buying public were confused about the purpose of The Cat? Sales suffered, and with it, the desire of Mercury to push the performance envelope.

Bold, Beautiful, and Aggressive Styling

More growth. Model year 1970 did not see any major changes, because it was only the second year of the second generation. A facelift consisting of modifying the grille, longer again, new paint offerings, as well as minor interior changes, were the extent of modifications. From a mechanical standpoint, the new year was carried over from 1969 relatively intact. Some engine changes, minor ones really,

One of 444 built, the 1970 Eliminator featured a new split grille and a third grille was inserted between the hood extensions. Equipped with the 428CJ and Ram Air, the 335 manufacturer horsepower rating was lower than actual output.

minimized buyer confusion. The "bold, sportier flair" grille center grew 2.3 inches, with the vertical pattern echoing the 1967 model. The Eliminator option carried over the black painted grille and taillight trim. A handful of "competition" colors were introduced: yellow, blue, gold, green, and orange. The Eliminator also included a black hood tape that swept into the black hood scoop, a passenger-side "racing" mirror, and a black front spoiler. The rear spoiler was color-keyed to the body color and wore a black "Eliminator" tape stripe. In midyear, a trio of new colors were added: red, deep gold metallic, and pastel blue.

Convertibles were in the mix, but the buying public was showing disinterest in ragtops. While overall sales of the Cougar continued to slip to 72,363 in 1970, less than half the introductory year levels (150,893 in 1967), the sales levels of convertibles really plummeted. In 1969, 5,796 standard convertibles and 4,024 XR-7 drop-tops were sold. In 1970, a total of 2,322 standard convertibles and only 1,977 XR-7 soft-tops were sold. Instantly, owners had a rare collector's item.

The interior was massaged as well. Trim styles included "upbeat" patterns of black or medium-brown hound's-tooth check cloth-and-vinyl. High-back bucket seats were made standard across the entire model line. Column-mounted anti-theft ignition, steering, and transmission lock were also added to the line. The three Cougar models were differentiated at the instrument panel. Simulated rosewood appliqué meant that you were in the standard, two-door hardtop model. The upscale XR-7 model was graced with simulated burled walnut appliqués and an XR-7 emblem. The Eliminator's basic black instrument panel maintained a more utilitarian look with a 6,000 rpm tachometer unless the buyer ordered the Boss 302 engine option. Then the driver faced an 8,000 rpm unit. Both the XR-7 and Eliminator models carried a full complement of gauges, including alternator, temperature, and oil pressure gauges as well as a tachometer. Even the steering wheels were different. The Eliminator and base trim level Cougars had the steering wheel center pad–actuated horn. But the XR-7, as well as the regular Cougar so optioned, came with the "Luxury" rim-blow steering wheel, a small rubber trim piece following the circumference of the wheel. When it was squeezed, the horn sounded.

With front and rear spoilers and the bolder 1970 graphics package, the 1970 Cougar Eliminator in Competition Orange is hard to miss. The graphics package features a tape stripe running the entire length of the vehicle as opposed to the spear stripe used on the previous year

Chassis and Suspension for High-Peformance Handling

This was to be the last year of the true high-performance Cougar. Government-mandated emission standards were coming down the pike, and these standards would emasculate the musclecar. But in the 1970 model year, the tone was still how much horsepower could be stuffed under the long, feline hood. For this year, Mercury used "New Cougar Eliminator. Spoilers hold it down. Nothing holds it back" for advertising purposes. The base suspension was elevated to the level of the previous year's "Competition Package." For 1970, the Competition Handling Package was standard on the Eliminator and optional on Cougar and Cougar XR-7. Handling, riding comfort, and traction were elevated to a new level. The Cougars equipped with the 351 engine had spring rates that were 14 percent stiffer in the front and 21 percent stiffer in the rear. The Cougar's front suspension was conventional for its day. A short-arm, long-arm design, it housed the shock absorber within the coil spring. A curved compliance strut with a soft front bushing at the lower front suspension anchor point helped deliver the "Mercury" ride, while helping to numb whatever road feel might find its way to the steering wheel. Speaking of steering, the power steering option quickened the steering ratio by 22 percent compared to the manual steering. With finger-tip effort, it took three turns to go from lock to lock. The front stabilizer bar grew to a diameter of .95 inch, and a "competition" suspension package included a .50-inch stabilizer bar in the rear to help control understeer. The rear four-leaf semielliptical spring rear suspension was typical for the era. Springs measuring 59 inches long and 2 inches wide were attached to the body using tuned bushings and Iso-Clamp attachments between the springs and the rear axle.

Any trim level of Cougar could be armed with the 428 engine, which was equipped with a 5/8-inch stabilizer bar hooked to the rear suspension. The shock absorbers had 1 3/16-inch diameter bores and provided 70 percent more control over the 1-inch bores on the standard shocks. The 428CJ Cougar Eliminator could be ordered with the Drag Pak option, consisting of an engine cooler, special heavy-duty engine parts, a 3.9:1 Traction Lok differential, and ram air induction, which made it Super Cobra Jet.

The Boss 302–equipped Eliminator was available with the Super Drag Pak option, which added a 4.30:1 Detroit Locker differential to the aforementioned Drag Pak goodies. Any Cougar equipped with a four-speed transmission was automatically fitted with a Hurst shifter. The exhaust system was changed and featured a true dual exhaust rather than a single transverse muffler with dual outlets. The wheel on the base and XR-7 Cougars was 14x6 inches, while the width of the wheel on the Eliminator was increased to 7 inches. Most of these were the five spoke–styled steel wheel. Belted-ply tires were used across the entire Cougar range, from E78x14 blackwalls to F70x14 blackwall traction tire with raised white letters. At curb weight of 3,476 pounds, the Eliminator was not a featherweight contender. Optional power-assisted front disc brakes helped bring the sporty Mercury out of the warp speeds the large engines could generate.

In the competition of the musclecar wars, the yardstick used was brutal acceleration. *Car Life* took a Boss 302 Eliminator to the drag strip and pulled a 0-60 miles per hour time of 7.6 seconds. The same car covered the quarter-mile in 15.8 seconds and crossed the finish line at 90 miles per hour. And while the Cougar Eliminator did not need to apologize to anyone, its luxury-mandated bulk put it at a performance disadvantage. With rivals such as the

There was not much clearance between the aluminum valve covers and the shock towers. The chassis change in 1969 was engineered to allow this size engine to be shoehorned in. The opening in the bottom of the hood fit against the rubber seal on top of the air cleaner.

Next Page
Black was a rare color on 1970 Cougar Eliminators, and it gave the Merc a stealthy, sexy look that fit this unique musclecar well. For 1970, the 351-4V Cleveland engine joined the Boss 302, 390, and 428 powerplants.

In 1970, only 450 Eliminators were equipped with the potent Boss 302 engine. The Boss 302 was a truly special engine. The engine had oversized valves, four-bolt main bearing caps, and 10.6:1 compression ratio. There were no external indications that this race-honed powerplant was lurking under the hood.

Camaro/Firebird, Dodge Challenger T/A, and its corporate cousin, the Mustang, it wasn't the fastest car in its class. It was on a more equal footing with vehicles such as the Buick Gran Sport, Oldsmobile 4-4-2, and Pontiac GTO Judge.

Changing Times, Changing Market, Changed Cougar

The muscle intermediate market was evaporating. Most manufacturers were starting to steer their mid-line offerings toward the boulevard cruiser rather than the boulevard crusher. Sales of performance cars had been in a decline, and the government regulations in the pipeline guaranteed that fire-breathing vehicles like the strong Cougars would become a thing of the past. With a sense of inevitability, powertrain options shrank. Oh, what a difference a couple of years make.

This Boss 302 engine did not tax the confines of the engine compartment like the 428CJ. A high-revving, solid-lifter mill, it put out considerably more than its factory-rated 290 horsepower. The acceleration from this small block wonder was nothing short of stunning. It could scamper from 0 to 60 miles per hour in about 7 seconds.

The Boss 302 engine, a mid-1969 release, had mechanical valve lifters on independently mounted rocker arms and a forged steel crankshaft that provided high strength with minimum weight. *Ford Motor Company*

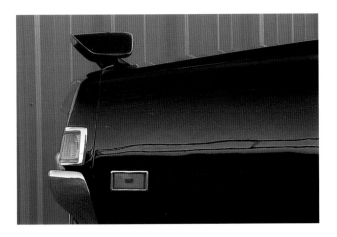

The decorative rear bumper was not really expected to seriously protect the vehicle, but it was a valuable styling tool. Eliminator script could be easily overlooked with the rare black paint.

The driver of a Boss 302-equipped Cougar Eliminator faced an 8,000 rpm tachometer and handled gear changes with the factory Hurst shifter. The Comfort-Weave upholstery ensured plush comfort and was part of Mercury's pleasing ride, every time.

The 1969–1970 Eliminators featured a dropped styling line similar to the 1969 Buick Skylark. Triple sequential taillights were still being used on all Cougar models, and the side-marker lamp lenses had the Ford logo imprinted on them.

Neither nimble handling nor controlling weight was reflected as the 1971 Cougar grew in size and weight. It was larger, again, partly to make room for the 429 Cobra Jet engine option. But in virtually all dimensions, bigger was the trend. The wheelbase was stretched 1 inch to 112.1 inches. Although the overall vehicle length was relativity unchanged, the car had a much larger appearance. The trademark hidden headlights were gone, replaced by four lamps set in a recessed grille. The center of the front end was dominated by a massive grille that resembled a chromed radiator. The rear window treatment was lifted off of

the Marauder X-100. And with options like a wrapover Landau roof treatment on the XR-7, the focus of Mercury's efforts to market the Cat was shifting to cruise mode.

For 1971, three models of Cougar were offered, the base model, the XR-7, and the new GT appearance option, taking the place of the dropped Eliminator as an option with the base model hardtops. Powerplants were a bit thinner than before, with a two-barrel 351 Windsor the standard engine on all models. The 351W put out 240 horsepower at 4,600 rpm, had a 9.0:1 compression mill, and pushed 350 foot-pounds of torque to the ground at 2,600 rpm. For people wanting to upgrade the engine compartment, the 351 Cleveland was available; its 10.7:1 compression ratio produced 285 horses at 5,400 rpm and 370 foot-pounds of torque.

The most powerful engine choice was the hydraulic-liftered Cobra Jet 429, which boasted of 370 horsepower at 5,400 rpm with 450 foot-

The long hood and short rear deck were musclecar requirements. The stance exuded a purposeful air, even at rest. A 111.1-inch wheelbase and a Mercury-tuned suspension contributed to a comfortable ride.

From the press release dated February 3, 1970: "The houndstooth Cougar – a car dressed like a fashionplate – proved the inspiration for women's fashion designer Pauline Trigere's new swashbuckling houndstooth cape. For a fashion-coordinated look, two great color schemes are available – medium brown check, as shown on both cape and Cougar, and black check." *Ford Motor Company*

pounds of torque at 3,400 rpm. Its 11.3:1 compression required premium fuel flowing through the Rochester Quadrajet carb. If the optional Ram-Air was installed, power increased to 375 ponies. The Drag Pak, which included mechanical lifters, adjustable rocker arms, and a 780-cfm Holley carburetor, turned the 429 CJ into a 429 SCJ. The differential was filled with either a 3.91 or 4.30 Traction Lok rear axle gear set or a 4.11 gear with a Detroit Loker rear end. The GT option could be ordered with any Cougar, regardless of engine choice. It equipped the Cat with a hood scoop, competition suspension, a tachometer, dual color-keyed "racing" mirrors, and a higher numerical rear axle ratio.

Midyear 1971, the 351C was slapped with a decreased compression ratio, and with that, less power. Five horses were cut from the output, but to make up for the loss, the engine was given a new name, the 351 Cobra Jet. This was indicative of the future of the performance line. Marketing now had the responsibility of compensating for the evaporation of the musclecar.

Dawn of a New Age

Cougar and Capri 1972–1989

8

The landscape had changed. Previous generations of musclecars roared past the paltry offerings built from 1972 on. In essence, high performance went into steep decline and eventually died. Mercury, like many of its competitors, was walking the path to economic ruination. The government was a major contributor to the demise of brute power as were escalating insurance costs. The 1973 oil crisis was the stake in the heart of the musclecar market. The resulting corporate response was to emasculate the engines of old that had made a name for Mercury. There was no doubt that cleaner air was an issue that needed attention, but scores of new regulations from Washington simply pulled the carpet out from the performance market. Detroit was caught in dire straits. Emission controls sapped power, heavy safety equipment added weight, and automotive electronics were in their infancy, so old-school technology was being called upon to solve new challenges. Sometimes the results were less than satisfactory. The marketplace was now home to the shadows of muscle. The Cougar was slowly being de-clawed.

Cougar's Downward Performance Spiral

When the 1972 Cougar hit the marketplace the price was reduced. This was in response to a disturbing trend that Mercury had noticed: shrinking sales. People were not flocking to sign on the dotted line in the numbers that the accountants liked, and a base coupe went for $3,016. And what did that buy? Styling like the year before but with less under the hood. The standard engine was the 351 Windsor, which had 8.3:1 compression and two-barrel carburetor producing a paltry 163 horsepower out at

The Fox-chassis Mercury Capri hit the streets in 1979, and it used the same 302-ci V-8, transmission, and suspension as its sister—the Mustang 5.0. But the Capri had a personality all its own: it used a different rear window, square grille, and square-jawed spoiler. The Capir never carved out a large enough following, and this 1985 RS model was the last of its generation. *Ford Motor Company*

The 1971 Cougar grew, losing a bit of the sleekness it had possessed. The Cat tipped the scales at a touch over 3,400 pounds and the Standard Cougar convertibles were somewhat rare, only 1,723 being built. *Ford Motor Company*

Hurst supplied the shifter on manual transmission–equipped Cougars. Standard Cougars did not have the dash toggle switches and woodgrain appliqué of the XR-7. *Ford Motor Company*

3,800 rpm. In 1972, auto manufactures started releasing engine outputs in SAE "net" figures, which were lower than the old system. These numbers were more "real-world" realistic. The top engine in the lineup was a CJ351-ci Cleveland engine equipped with a four-barrel carburetor, hydraulic lifters, and 9.2:1 compression. It was rated at 266 horsepower at 5,400 rpm and 301 foot-pounds of torque at 3,600 rpm. Between the two engines was a third, the standard engine with the four-barrel from the CJ351 and 8.8:1 compression. It was reported to produce 262 horsepower at 3,800 rpm and 301 foot-pounds of torque with its special dual "Quiet Exhaust" system. The big-block engines were dropped for '72. The big-block's return would take a couple of years, and when it happened, it was not a triumphant return of horsepower.

The Cougar sales trend continued downward through the 1972 model year, yet the rest of the division was doing well. William P. Benton, general manager of the Lincoln-Mercury Division, said "Despite serious challenges to many areas of our business, sales have never been better, and the future has never been brighter." The division as a whole saw a 30 percent increase in total sales from the prior year, but the

The exposed headlights and massive center grille on the 1972 Cougar were a sign that Lincoln design was playing an increasing role in the Cat's style. This was the first year that a big-block was not available. *Ford Motor Company*

Cougar continued to be a mediocre seller as only 53,702 Cats went to a good home. So how did Mercury address this challenge? They left the car alone.

In 1973, government mandated new bumper laws. The front bumper was required to withstand a 5 miles per hour impact, and the rear had to take a 2.5 miles per hour hit. Putting these chrome-plated railroad ties on the vehicle did nothing for looks, handling, or mileage. Two body styles were available, or more properly, two levels of trim—the standard Cougar and the upscale XR-7. Adding insult to injury, the engine compartment was now home to only two engines, both of 351-ci displacement. The standard two-barrel engine developed only 168 horsepower at 4,000 rpm, and just 256 foot-pounds of torque at 2,400 rpm, the lowest figure in Cougar's history to that point. In a vehicle that tipped the scale at about 3,400 pounds, acceleration was less than inspiring. The optional 351 Cleveland four-barrel engine put 264 ponies under at 4,800 rpm with 314 foot-pounds of torque. These pathetic numbers were a sad commentary on the state of performance at Mercury. Yet sales of Cougars increased in model year 1973, with 60,628 units leaving dealer lots.

The Last Original Cougar

It was a first for Cougar—one model. And in the finest Lincoln-Mercury tradition it was the top level, $5,218 XR-7. The 1973 model had been the last Cougar built on the original floor pan. The 1974 Cougar was a new vehicle, having moved up in product placement. A Cougar advertisement of the time boasted that "In fact Cougar is the only new choice among the mid-size personal luxury cars." Nothing in that ad about performance. The Montego, the workhorse intermediate at Mercury, had to fill the void left by the new Cougar. The Mustang had morphed into the Mustang II, and there was no way that Mercury was going to produce a compact sports car.

The styling was similar to the year before; the car was just a bit larger and had more of a Lincoln touch, including C-pillar opera windows, an upright hood ornament, and an inside hood release to protect those valuable performance engine parts. The Cougar rode on a 114-inch wheelbase and had an overall length of 215.5 inches. The 4,255-pound car was propelled by one of four engines: the base unit, a 351, a 400-ci mill, and the 460. The standard unit was the 351 Windsor with 8.0:1 compression and a two-barrel carburetor that produced 168 horsepower at 4,000 rpm. The 400, a 351 with a higher deck height and larger crankshaft bearings, had a two-barrel and pumped out a stirring 170 horsepower. Moving higher on the option sheet resulted in the four-barrel 351 Cleveland engine that in fact had the highest power rating of the model year with 255 horses. The last offering was the big Ford 460, which FoMoCo was using in their largest passenger vehicles. This full-sized machine was a 8.0:1 compression, four-barreled engine putting out 220 horsepower with 342 foot-pounds of torque. That was a lot of low-end grunt. Attached to all of these engines was a Select Shift Cruise-O-Matic automatic transmission, the only

choice. At least a high- performance axle and Traction-Lok differential were available. But the public was buying. Cougar sales were an improvement at 91,670 cars sold.

The 1975 Cougar contained one new styling change. It had a pair of air openings in the lower grille below the front bumper. That was the extent of the design department's stylistic efforts. However, plenty of new options were offered, such as the space saver spare tire, power moonroof , security lock group, and a heavy-duty electrical system to power all the accessories. What little horsepower the engines were putting out seemed to go to the alternator, which did not leave much for the tires.

The base engine was the trusty 8.0:1 compression 351W or the 351M (the 351 Modified was a destroked 400), which sat under a two-barrel carburetor and spit out only 148 horsepower at 3,800 rpm. The two-barrel 158-horsepower 400 and four-barrel 216-horsepower 460 were still available, but power was taking a downward spiral. Emission controls were sapping the life out of engines that only a few years ago had ruled the streets. But gasoline prices were climbing like a homesick angel, and the last thing the public needed was a car that saw single digit mileage. Economy cars were selling like hotcakes, and the word was out that the large car was doomed and on its way out. Evidently, 62,989 people did not get the word in 1975 and took home a new Cougar instead.

The Cougar's Spirit of '76

Disco was in full swing, and flashy was the accepted look. And what better way to arrive than in the stylish XR-7, a rolling example of American Baroque. The Cougar had become a downscaled Lincoln and was much more similar to the Thunderbird than ever before. Any pretense of performance was gone, replaced by the ability to waft to the opera in a rolling isolation chamber. To hold down costs, several items that had been standard were now options, but at least they were still available. Unfortunately, the same could not be said for the engine compartment. The base engine was the 351 Windsor, an 8.0:1 compression ratio being fed through a Ford two-barrel carburetor. With 152 horsepower at 3,800 rpm and 274 foot-pounds at 1,600 rpm, this engine was a low-rpm cruiser. If the buyer yearned for more power, two optional engines were available. The first was the 400-ci mill, putting out 180 horsepower at 3,800 rpm and 336 foot-pounds of torque at 1,800 rpm. Equipped with an 8.0:1 compression ratio and the same two-barrel carb as the base engine, it was a $93 option. The other option was the big 460 engine. Like all the others it was fitted with hydraulic lifters. It too sported an 8.0:1 compression ratio that produced only 202 horsepower at 3,800 rpm with 352-foot pounds of torque at an extremely low 1,600 rpm. The power curves of these engines were targeted for effortless boulevard cruising, not for spirited driving. But Mercury was tapping into a valid market, as 83,765 Cougars were snapped out of dealerships. This must have been encouraging to the executives at Mercury as 1977 saw a new Cat prowling the streets.

A new body shell took the Cougar to another level of luxury. The Montego name evaporated, and the former Montegos became Cougars. The full line of Montegos, including four-door sedans and station wagons, now had Cougar emblems. The XR-7 was fitted with a different trunk lid to differentiate it from the standard Cougars. The large grille was set off by the four exposed rectangular headlights, and it weighed in at a hefty 3,909 pounds. With its spare tire bulge and vertical taillights, it really looked like a sibling of the Lincoln, but it sure didn't have Lincoln-like power under the hood. The 302-ci standard engine was equipped with a 8.4:1 compression ratio and a two-barrel Motorcraft 2150 carburetor. It produced a sad 130 horsepower at 3,400 rpm.

The A-shell platform, shared with the Ford Thunderbird, was the basis for the 1977 Cougar. The largest engine available was the 400-ci mill, putting out a whopping 173 horsepower thanks to emission regulations. *Ford Motor Company*

For 1978, the XR-7 was pretty much carried over intact. Changes were limited to paint and options. At least the station wagon was not a Cougar anymore. The Cougar could be outfitted with a Midnight/Chamois Decor Option. The highlight of this package was the Chamois-colored padded vinyl deck lid appliqué that covered the spare tire bulge. There were 166,508 XR-7s bought in model year 1978. The 302, standard again, enjoyed a power gain—of five horsepower, to 135 at 3,400 rpm. The 351s were still available, making 145 horsepower at 3,400 and 273 foot-pounds at 1,800 rpm. This was the last year for the 400-ci engine, fuel economy pressures pushing the engine out of the showroom. But in its last year, it made 160 horsepower at 3,800 rpm. Mercury was pleased indeed that a record 166,508 XR-7s hit the road. With that kind of buyer response, the end of the 1970s was also the end of the huge Cougar. But performance glimmers were seen on the horizon.

Except for the discontinuation of the 400 engine, the 1979 Cougars were just like the 1978s. A 302-ci V-8 was still the base engine, and a 351W with 135 horsepower, and 351M with 151 horsepower were still available. But the big news for Mercury performance was the release of the "new" Capri.

Sports Car from a Distant Shore— Mercury Capri

In May 1970, Lincoln-Mercury started selling the 1971 Capri, a creation of Ford Germany. Equipped with a small, 97-ci engine making 75 horsepower, it was Mercury's attempt to capture the sporty market. With the musclecars a thing of the past, Mercury wanted to continue to attract a younger clientele. Thus the Capri was brought over until 1978. Buyers of this "import" stayed away in droves sending the "sports" car back across the pond. In 1979, an all-new Capri was introduced, based on the Mustang. It came only as a 2,645-pound three-door hatchback, and the 100.4-inch wheelbase vehicle was available as a turbo-charged four-cylinder, a 2.8-liter V-6, and a 5.0-liter V-8. The 5.0 engine was the Mercury Man's version of muscle, and it flexed at 129 horsepower at 3,400 rpm, while the torque rating of 223 foot-pounds was measured at 1,600 rpm. While it wouldn't threaten a ten-year-old Cougar, it was a real improvement over the imported Mercury. And in its eight-year life span, the re-badged Mustang would be the closest thing to performance resurrection.

1974 MERCURY COUGAR XR-7

In size, this new breed of Cougar is like Grand Prix and Monte Carlo. In every other way, it's like nobody else's car.

You're looking at the all new Cougar for '74. It's more than a new car. It's moved up one whole class. In fact Cougar is the only new choice among the mid-size personal luxury cars.

There's new styling, inside and out. New dash with tachometer and hooded gauges mounted in deeply padded vinyl. Elegant new opera window. Distinctive new Landau roof. Steel-belted radials. All standard. There's power steering. And front disc brakes, automatic floor shift and bucket seats, also

standard. Plus the same type suspension system as Lincoln-Mercury's most expensive luxury car. Other features shown are optional.

And along with Cougar's new size class comes a whole new class of comfort for you. Because we felt this much luxury deserved a little more room.

MERCURY COUGAR

LINCOLN-MERCURY DIVISION *Ford*

The comparisons to competitors showed the direction that Mercury wanted to take the Cougar in 1974. From its battering ram bumper, and upright hood ornament to the C-pillar opera windows, Mercury's personal car had evolved into a boulevard cruiser. *Ford Motor Company*

Of course, optional engines were waiting to be ordered. The 8.3:1 compression 351W was still around, cranking out 149 horses at 3,200 rpm, and the 351M was still on the option sheet, replete with 8.0:1 compression, which helped produce a lofty 161 horsepower at 3,600 rpm. If yet more power was wanted, the 400-ci engine could be installed. This top-line engine put out 173 horsepower at 3,800 rpm from 8.0:1 compression. The torque reading of 328 foot-pounds at 1,600 rpm was as good as it got in 1977. Yet sales for the XR-7 increased to 124,799 cars. This was a signal to Mercury to brush on the glitz.

When the Capri debuted, it sold well in its first year—110,144 units. It could provide impressive performance if the right option box was checked. A turbocharged four-cylinder 2.3- liter mill producing 140 horsepower at 4,800 rpm was one of four engines offered. It had 9.0:1 compression and two-barrel Holley 6500 carburetor, iron head and block, and hydraulic lifters handling the valves. The other strong engine for the Capri was the trusty 302 5.0 liter. This two-barrel– equipped iron engine made 129 horsepower at 3,400 rpm from 8.4:1 compression, and 223 foot-pounds of torque at 1,600 revs. In 1980, this engine was downsized to 255-ci displacement, fuel economy needs being a high-profile concern at the time. Sales downsized a bit, too, with 79,984 units sold. With the 255's compression at 8.8:1, it put out 115 horsepower at 3,800 rpm and 191 foot-pounds of torque at 2,200 rpm. The 2.3-liter turbo-four was still the top engine in the lineup, but Mercury was having problems with head gasket failure and the turbo reliability. The turbo wasn't intercooled and suffered failure from heat buildup. Thus, the engine did not return for 1981, and the top performance engine option was the 255-ci V-8. The horsepower remained the same, while torque was up to 195 foot-pounds at 2,000 rpm.

Sales of Capris were a bit soft in 1982, with 36,134 vehicles sold, but Mercury had high hopes that the infusion of horsepower would bring buyers back to the muscle Mercury. The 5.0-liter engine returned that year, equipped with an automatic transmission only. This 8.3:1 compression ratio engine was rated at 157 horsepower at 4,200 rpm and 240 foot-pounds of torque at 2,400 revs. When this engine was ordered, the Capri also came equipped with power steering, traction bars, and a Traction-Lok rear axle.

As technology was making advances in electronics, metallurgy, and computer-assisted design, performance was realizing some real gains. This led to a return in 1983 of that historic indicator of power, the four-barrel carburetor. Installed on the 302-ci (5.0) engine, the Holley 4180 helped bring 175 horses to the table at 4,000 rpm. In the middle of the year, a Borg-Warner five-speed manual transmission was made available in the H.O. model to assist in maximum acceleration. There were 25,119 Capris sold in model year 1983. The turbo-four was brought back for 1983 and 1984, but buyers opted for additional cubic inches more often than not. And in 1984, there were two reasons to choose other than the turbo. The 5.0 engine came in four-barrel or fuel-injected config-

The vertical ribbed grille, a Mercury tradition, dominated the wide front end of the 1978 Cougar. While the biggest engine was the 153- horsepower, 400-ci V-8, the emphasis was on luxury. *Ford Motor Company*

urations. The injected engine had a compression ratio of 8.3:1, allowing it to produce 165 horsepower at 4,000 rpm and 245 foot-pounds of torque at 2,200 rpm. The Holley 4180C-equipped 5.0 also enjoyed 8.3:1 compression, and on the RS model, tubular exhaust headers, 2.5-inch exhaust and dual outlets, and a high-lift camshaft covered by cast-aluminum rocker arm covers. The rear axle was fitted with four shock absorbers to try to control the axle hop that these vehicles suffered from under hard acceleration.

Performance surged forward in 1985. The turbo era had reached the end of its road, never to return. The 5.0-liter engine was gone through, and after the hotter camshaft and roller tappets were installed, the engine was rated at 210 horsepower at 4,400 rpm. The 8.3:1 compression and Holley four-barrel carburetor helped contribute to the 270 foot-pounds of torque that was fed through the Goodyear Eagle tires mounted on aluminum 15x7 wheels. Unfortunately, only 16,829 Capris were sold in 1985, which was down from the 17,114 units sold in 1984. The 1986 Capri contained the last carbureted 5.0 engine. Sequential fuel injection, new rings, and tuned intake/exhaust manifolds combined to make 200 horsepower at 4,000 rpm and 285 foot-pounds at 3,000 rpm. The compression ratio was raised to 9.2:1, and while the horsepower had dropped slightly, the Capri's drivability was greatly improved. With 1986 model sales of only 13,358, Mercury pulled the plug on the Capri. Dearborn would recycle the name a few years later on the Australian built two-seat convertible, but it would suffer a similar fate. Mercury knows how to market mid- and full-sized vehicles, but niche vehicles have lacked public acceptance.

The Epitome of a Family Coupe—Cougar

In 1978, to take the place of the defunct Comet, Mercury had introduced a new car called the Zephyr, which was a Mercury clone of the Ford Fairmont. The mundane front-engined, rear-wheel drive vehicle proved reliable. In 1980, Mercury released a new Cougar, built on the Zephyr floor pan. It had a 108.4-inch wheelbase, and came in one flavor, the Cougar XR-7 Sports Coupe. The styling was even more cubist than its predecessor. From its massive, upright, finned grille to the huge C-pillar, it would never be mistaken for anything from Europe. But Mercury engineers had started moving in the right direction with the new car. They had shaved 700 pounds off of the weight. It featured improved handling, aerodynamics, and Recaro bucket seats. All of the traditional Cougar touches, such as woodgrain appliqué, power everything, and a vinyl roof were offered to placate anxious buyers. But the powertrain had seen a bit of a change.

The 302-ci engine was no longer the standard engine. The base mill was the 255-ci (4.1-liter) V-8, putting out 115 horsepower at 3,800 rpm. The optional 302-ci engine had a 8.4:1 compression ratio, and the Motorcraft 7200VV two-barrel carburetor coaxed 140 horses out at 3,400 rpm. But the public seemed underwhelmed with the vehicle. In 1979, Mercury sold 163,716 XR-7s, which was the last year of the big Cougars. The new downsized '80 versions only saw 58,028 units sold.

The Mercury Monarch, a four-door mid-sized sedan, was given the ax for 1981, so the Cougar picked up a sedan in the lineup. Again. And for the first time since the XR-7 was released in 1967, a V-8 was not the standard engine. That honor now belonged to the venerable 200-ci straight six-cylinder engine.

The 302 engine, the optional mill in the XR-7, was shorn of even more of that pesky horsepower in 1981. With an 8.4:1 compression and a two-barrel carburetor, it pumped out 130 horsepower at 3,400 rpm. The XR-7 found only 37,275 people who would write the check. Not good news for the Mercury executives [again]. They gave the Cougar one more year of life-support.

When 1982 rolled around, the biggest engine was the 255-ci (4.2-liter) V-8, which put out an embarrassing 120 horsepower at 3,400 rpm. Only 16,867 XR-7s were sold, and that was all she wrote for the Zephyr-based Sports Coupe. The Cougar name had sunk about as low as it could, and drastic action would be needed to start the name back to credibility. Fortunately, Mercury showrooms had that chance in model year 1983, when a courageous Ford Motor Company put it all on the line with an entirely new look for their Mercury division. Aero…

The Third Generation Cougar

It looked as different as night and day. The 1983 Cougar shared the running gear and floor pan of its Ford cousin, the Thunderbird. The new Cougar had a formal, upright rear window which was distinctly different from the T-Bird's sloped roofline that gently curved all the way to the taillights. But the front end sliced through the air, unlike its predecessor's blunt, barn-door school of aerodynamics. The coefficient of drag on the new Cougar was 0.40, a very good number for the era. Riding on a 104-inch wheelbase, the new Cat got a reading in the neighborhood of 2,900 pounds. An XR-7 wasn't offered in 1983, just Cougar. The base engine was a gutless 232-ci, (3.8 liter) 110-horsepower V-6. A two-barrel Motorcraft 2150 or 7200VV carburetor breathed into the 8.65:1 cylinders. But at least the option sheet had a V-8 for the Cougar. With a compression ratio of 8.4:1 and fuel injection, its 130 horsepower at 3,200 rpm and of 240 foot-pounds of torque at 2,000 rpm was somewhat pathetic. At least the foot-pounds rating was an improvement over the prior year.

However, help was on the horizon. The amount of electronics was increasing every year, and with onboard computers starting to be used under the hood, emissions could be closely monitored and the engine could adjust various engine operating systems to maintain desired operating parameters. And the public's response to the new body? There were 75,743 Cougars sold in 1983. The new look was accepted by the public, but all of a sudden everything else looked very dated.

The largest engine offered in 1979 was the 351M, rated at 151 horsepower. During the last year in which the Cougar was built on the T-Bird platform, the vehicle weight in the neighborhood of 3,800 pounds did nothing for performance. *Ford Motor Company*

The momentum that started in 1983 kept building the next year. The XR-7 name was back, fitted with a turbocharged 140-ci four-cylinder engine, blowing out 145 horsepower at 4,600 rpm. Buyers that wanted a less-stressed powerplant could opt for the fuel-injected 302 engine with 140 horses at 3,200 rpm and 250 foot-pounds at 1,600 rpm. All engines were now fitted with EEC-IV electronic controls to monitor vital signs. Cougar popularity and sales were increasing. Total Cougar sales in 1984 were 131,190, which was a significant jump from the prior year.

A minor freshening of the front and rear was performed for 1985. And even better news was that a bit of a power increase had found its way under the hood. The four-cylinder turbo was still the standard engine with the XR-7, yet the 5.0 liter was alive and kicking. The Turbo raised its output to 155 horsepower at 4,600 rpm. And the 5.0-liter V-8 had not been ignored, and it was now putting out 140 horsepower at 3,200 rpm. Fuel injection helped the engine to reduce smog, improve mileage, and boost power. Technology was finally catching up with the demands of D.C. And the die was cast; performance

helped to sell cars. With sales of 117,274 Cougars in model year 1985, Mercury was breathing a bit easier about its Cat.

Shades of High Performance

The Cougar in 1986 was coming into its stride. Mercury was moving the horsepower levels in the proper direction, with acceptable fuel economy and legal emissions. Fuel injection had replaced carburetors, and the use of computers under the hood had increased. The formal styling of the Cougar had some detractors, but sales were climbing. In fact, for the '86 model year, Cougar was the top selling vehicle in the Mercury catalog with 135,909 cars rolling off the showroom. Performance might have had something to do with it, as the motive power was rising. The 3,000-pound XR-7 used a turbo-charged four-cylinder as the standard engine, but it was still rated at 155 horsepower at 4,600 rpm, provided the five-speed manual transmission was installed. Fitting the Cougar with the optional automatic resulted in a drop to 145 horses at 4,400. Torque also had a split in numbers depending on transmission choice. But for lovers of old-fashioned grunt, the 5.0-liter engine was the way

Introduced in 1979, the Mercury Capri was a badge-engineered Ford Mustang, built on the same "Fox" platform. Sold only as a three-door hatchback, the top-line Capri RS used a 302-ci engine that only produced 129 horsepower. *Ford Motor Company*

to go. Sequential multi-port fuel injection was installed on the V-8, with positive effects on power generation. Compression was raised to 8.9:1, and horsepower was now 150 at 3,200 rpm. Torque went up as well, the iron block churning out 270 foot-pounds at 2,000 revs. This was the last year of the turbo-four in the Cougar, as the 5.0-liter engine was an easier path to power.

The base engine in the Cougar for 1987 was a 90-degree, V-6 iron-block, aluminum-headed engine displacing 232 ci (3.8 liters). This mill made 120 horsepower at 3,600 rpm with a 8.7:1 compression ratio. But there was no manual transmission available in the Cougar; all shifting was being done by a four-speed overdrive automatic. Buyers needed to spend an additional $639 to put the 5.0 engine under the hood. The power outs didn't increase for 1987. And while it wasn't the biggest seller for Mercury that year, it had no need to apologize for the 104,528 units that were purchased.

Power took another jump upwards in 1988, with the base 3.2-liter engine now putting out 140 ponies

at 3,800 rpm, in part due to the raising of the compression ratio to 9.0:1. A balance shaft was installed to smooth out the V-6, and the fitting of multi-port fuel injection helped increase the torque rating to 215 foot-pounds at 2,400 rpm. But for those desiring more speed, the 5.0-liter engine was on the option sheet. For $14,855, the V-8 powered Cougar boasted of 155 horsepower at 3,400 rpm, while torque took a slight dip to 265 foot-pounds at 2,200 revolutions. Sales figures dropped to 119,162 for the Cougar. But this was the last year for this Cougar platform, as a new Ford Thunderbird was debuting in 1989, which meant that a new Cougar would also be released. The "old" Cougar had served Mercury well, and on its watch, performance had entered the vocabulary again.

A Rejuevenated Cougar for the 1990s

Mercury ended the 1980s on a high note with the December 26, 1988, introduction of the next Cougar, cousin to the Thunderbird again. The formal upright rear window was carried over, making the new Cougar easily identifiable as the Cat. Changes under the skin were significant, though, and for the better. The wheelbase was lengthened to 113 inches, and vehicle weight increased by about 400 pounds. But for the first time, the rear suspension was fully independent. Only two levels of Cougar were offered, the standard LS and the XR-7. The LS came with the 140-horsepower 3.8-liter V-6 attached to a four-speed autobox. The XR-7, retail priced at $19,650, was available with only one engine, a supercharged 3.8 V-6. No 5.0-liter V-8 was offered, but the new powerplant made up for the deletion. With a compression ratio of 8.2:1, the blown engine avoided the turbo lag problems of the past while kicking out 210 horsepower at 4,000 rpm and 315 foot-pounds of grunt. The engine was equipped with a five-speed transmission with a four-speed overdrive transmission as an option. The manual transmission XR-7 would run the quarter-mile in the mid-15s, which was superb performance for a vehicle that fully loaded was looking closely at 4,000 pounds. Only 4,780 XR-7s were constructed out of a total production run of 97,246. Shades of Cougar's early days, when a relatively few performance models generated ink far beyond their numbers, their aura covering the entire line.

The entry-level Capri in 1980 was equipped with a 2.3-liter four-cylinder engine, capable of producing a paltry 88 horsepower at 4,600 rpm. These drivers were safe from "exhibition-of-speed" tickets. *Ford Motor Company*

High-Performance Resurrection

Mercury 1990-2000

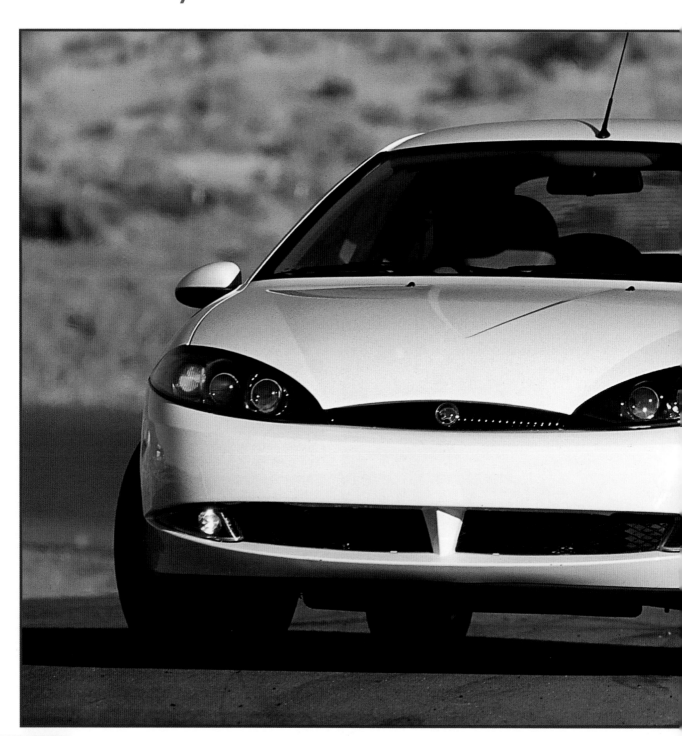

9

In the last decade of the second millennium, Mercury got back into a performance arena, though not like in the late 1960s. When horsepower was added to the production line-up, sales responded in a positive fashion. Mercury had a stronger Cougar in the lineup than they had had in many years. Electronics had found their way under the hood, allowing engineers to solve multiple challenges in a fraction of the time that it would have taken even ten years before. Both power and mileage were increased. So Mercury started the 1990s on a strong note with the supercharged Cougar XR-7 a visible indication that performance was not a dirty word in Dearborn.

The 1990 Cougar was essentially a carry-over model from '89, which was not a bad thing. The supercharged 3.8-liter V-6 still resided under the long hood and pumped out 210 horsepower at 4,000 rpm. The 8.2:1 compression engine squeezed out 315 foot-pounds at 2,600 revs, enough to smoke the 225/60VR16 tires. In the January 1989 issue of *Motor Trend*, the MT staff ran an automatic transmission–equipped XR-7 through its paces, coming up with a 0-60 time of 8.3 seconds. Brake performance was greatly enhanced with the addition of anti-lock brakes on the standard Cougar XR-7. Mercury sold 76,467 units, and out of that figure, only 4,129 Cougars were the XR-7. This was not the kind of trend that Mercury wanted to see that would justify putting horsepower in the hands of the public. But the profile of the Cougar buyer had changed from the vehicle's early years. The Cougar had been based on the Mustang, which imparted a certain amount of inherent sporty-ness. Cat buyers were getting the Mercury version of the Ford Thunderbird MN12 platform with a wheelbase of 113 inches.

The 1999 Cougar is pictured at Willow Springs Raceway during media evaluation. The crisp lines are representative of the New Edge design concept now in place at the Ford Motor Company studios.

The XR-7 was the only model of Cougar built in 1993. Though the base engine was the 3.8-liter V-6, an optional High Output 5.0-liter V-8 was offered, rated at 200 horsepower at 4,000 rpm. *Ford Motor Company*

Luxury was the priority, not smoking tires. In 1991, another performance engine was dropped into the engine compartment—the 5.0 liter V-8. Again.

The Cougar, built at Lorain, Ohio, was given a mild facelift for 1991 as well as the 5.0-liter HO (High Output) in the XR-7. Traditional Mercury Cougar styling cues such as a vertical ribbed grille and the snarling cat profile were carried over. This engine was standard in the XR-7 and was a $1,184 option in the base Cougar LS. The LS came with the standard 140 horsepower 3.8-liter V-6. From a performance standpoint, the 302-ci mill putting out 225 horses at 4,200 rpm with peak torque peak of 275 foot-pounds at 3,000 rpm was the only choice. The supercharged engine and the manual transmission were relegated to memory. The four-speed automatic overdrive transmission was standard on all Cougars. Its weight

lent a certain feeling of driver superiority over anything classified as a compact. The April 1991 issue of *Motor Trend* ran a complete road test that spoke highly of Mercury's personal luxury car. A 0-60 time of 9.4 seconds was achieved, and the quarter-mile was dispatched in 17.1 seconds with the finish line flashing by at 82.2 miles per hour. Cougar Eliminator or Cobra Jet drivers weren't going to quake in fear from the performance, but the Cougar was a much more refined vehicle than its predecessors from the musclecar era.

The same lineup returned in 1992 with the exception of a 25th Anniversary Cougar. In honor of the milestone, a limited-edition LS model was offered, complete with 5.0-liter HO engine, BBS wheels, monochromatic paint, and special trim. Powertrains were exactly the same as in the previous year.

The 1997 Cougar was the last year for a front-engine, rear-drive Cat. The optional 4.6-liter, overhead cam modular V-8 engine put out 205 horsepower at 4,250 rpm. There was no 1998 Cougar; instead, the new 1999 Cougar was released in early 1998. *Ford Motor Company*

Mercury was letting the Cougar go on cruise control. Total Cougar sales in model year 1992 were only 46,982 units, but Mercury did not feel that the nameplate was threatened.

Their faith in the public was justified, at least as far as sales were concerned. In 1993, Cougar sales jumped to 79,700 units. The big news for the year— LS model was dropped, and the Cougar underwent a name change. It was called Cougar XR-7 coupe, and its base engine was the 3.8-liter V-6 with sequential fuel injection that generated 140 horsepower at 3,800 rpm. The 5.0-liter V-8 was still available, but its power had been down-rated to 200 horses at 4,000 rpm, while torque remained the same at 275 foot-pounds at 3,000 rpm. Times were changing at Mercury, and the pushrod, 9.0:1 compression engine was put out to pasture at the end of model year 1993.

For 1994, the venerable 5.0-liter V-8 replaced an optional modular engine displacing 281 ci (4.6 liter). The Ford family's new high-performance engine featured a single overhead camshaft in each aluminum cylinder head, sequential port fuel injection that was 200 pounds lighter than the 5.0, and iron-block engine. It produced 205 horsepower at 4,500 rpm while the torque rating was 265 foot-pounds at 3,200 rpm from 9.0:1 compression. This vehicle tipped the scales at a reduced 3,726 pounds. Safety took a step

forward as driver and passenger airbags were now standard in the XR-7. The Cougar sold in respectable numbers for the 1994 model year, with 71,026 finding qualified buyers.

When 1995 rolled around, the Cougar series still consisted of one model, the XR-7 luxury coupe. The power output ratings for the base 3.8-liter and optional 4.6-liter engines remained as the 1994 models. The V-8–equipped XR-7 started at $17,305, but if the myriad of options available were ordered, the retail price rose to the mid-twenties. The Sports Appearance Group, complete with BBS wheels and a Nonfunctional Luggage Rack was an attractive option for Merc enthusiasts. But under the skin, the Cougars utilized rack-and-pinion steering, H-arm independent rear suspension, variable rate coil springs, and stabilizer bar. The Cougar provided excellent handling and ride quality for a car of its size. The standard P205/70R15 tires clung to the twisty corners with a surprising amount of aplomb. Nevertheless, sales were a little soft this year with 60,201 vehicles sold.

In 1996, the Cougar made strides on several fronts. Revised front and rear sheet metal offered a stylistic advancement. Engines were made more fuel-efficient and the scheduled maintenance intervals longer. The base engine was the 3.8-liter unit, and the 4.6-liter V-8 was a $1,130 option that cranked out

The new 1999 Cougar, equipped with the optional V-6 engine, overlooks the Willow Springs Raceway in Rosamond, California. The hatchback design is sculpted to appear like a conventional coupe.

205 horsepower at 4,250 rpm and 280 foot-pounds of torque at 3,000 rpm. The engine produced power peak at lower rpm and that translated into less interior noise. This was more in keeping with the luxury overtones that the XR-7 personified. Sales took a beating that year, and only 38,929 Cougars found their way out of the showroom. The platform was showing its age; after all it had been around since 1989. The basic shape had been seen by the public for the first time in 1983.

Special Edition 30th Anniversay Cougar

In 1967, few Mercury enthusiasts or executives would have anticipated a 30th anniversary Cougar. Surely not the criticsæthey had called the Mustang derivative a compromise car that tried to be a jack-of-all-trades and was a master of none. The Cougar started out a personal luxury car, and except for a spell as a sedan and station wagon, it has stayed a personal luxury car with some high-performance character. Mercury offered a 30th Anniversary Edition to commemorate the event, and badges and appliqués were the norm.

The special edition Cougar was fitted with cross-lace 16-inch wheels and sport shocks and was painted in Dark Toreador red. An engine that provided inspiring performance wasn't part of the package. The 4.6-

The ergonomically friendly interior of the 1999 Cougar is a comfortable place to put miles beneath the occupants. Dual front air bags are standard, and optional side air bags are located in the side of the seat backs. Available transmission includes a five-speed manual and a four-speed electronic overdrive automatic.

This is the hot performance engine for the new Mercurys. The optional Duratec 24-valve DOHC V-6 engine delivers 170 horsepower at 6,250 rpm, while delivering 19 miles per gallon in the city, 28 miles per gallon on the freeway with the manual transmission.

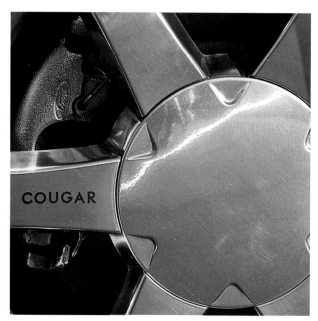

Optional four-channel, four-wheel disc brakes with ABS hide behind the 16-inch aluminum wheels and provide stunning brake performance for the 1999 Cougar. The tire fitted is a P215/50R16 blackwall.

liter engine was an option. In the XR-7, it would set a buyer back $18,960. This was one reason why only 35,267 were sold in 1997. In addition, the rear-wheel drive had fell out of favor with the a majority of the car-buying public. This was the last year of continuous production for the Cougar. The Cougar/Thunderbird platform was discontinued, and thus the Cougar became a part of Mercury's past, or did the Cat have another life?

Mercury's New Direction for Style and Design

Since the first production unit was built, the Cat has been manufactured in America. Although there was no 1998 Cougar, the next generation was unveiled at the North American International Auto Show on January 6, 1998, as a 1999 model. This vehicle was Ford Motor Company's first North American application of its New Edge design philosophy.

Mercury took advantage of the annual SEMA (Specialty Equipment Market Association) show in Las Vegas in late 1998 and unveiled a Cougar Eliminator concept car. Using off-the-shelf components, it was designed to show what an enthusiast owner could do to "enhance" the visual and performance characteristics of the 1999 Mercury Cougar. *Ford Motor Company*

Debuting at the annual SEMA show in Las Vegas was the Mercury Marauder concept car. Based on the Mercury Grand Marquis, its performance was improved with the addition of a Ford SVO supercharger onto the 4.6-liter SOHC fuel-injected V-8. The blacked-out grille is reminiscent of performance Mercurys from the late 1960s. *Ford Motor Company*

Former Ford Design chief Jack Telnack calls this styling direction the logical extension of the Aero look. Crisp lines combined with organic curves result in a mixture that looks like nothing else on the road.

After some preliminary work had been done, formal design work of the car started on January 1, 1995. Ford had spent a considerable amount of money developing the Mondeo platform for European use, and that structure was implemented in the United States as the Ford Contour/Mercury Mystique. Part of the "Ford 2000" program that was released in January 1995 called for effective utilization of existing resources. The New Edge was a styling departure from the rounded bulbous Aero Styling program. The New Edge featured sharp, bold lines and body creases. Ford wanted to develop a coupe on the platform that had been used as a sedan. That would help defray some of the tooling and development costs. The new vehicle had to come in both left- and right-hand drive models to allow overseas sales. From a styling standpoint, the goal was clear, as Jac Nasser, president, Ford Automotive

operations, put it. "Our challenge was to combine the athletic stance and agility of a sports car with the expressive elegance of a Mercury." Daryl Behmer, designer of the new Cougar, points out that the vehicle was actually designed in studios in both England and Cologne, Germany. The new Cougar, with a coefficient of drag of 0.32, shares the traditional vertical ribbing in the grille with its predecessors, but overall, it cuts new ground in Cougar design traditions. From the four-beam projector headlamps to the triangular taillight lenses, this new shape is meant to interest a new range of buyers. The target buyers are different from the prior generation of Cougars where an older group was enticed into the luxurious interior. The new Cat is intended for a group of 25- to 40-year-olds, about half of whom are female. Still, Mike Jennings, Cougar brand manager for Lincoln-Mercury Division, points out that buyers in the segment are "looking for style and image—regardless of age." The Cougar has been an image vehicle since its debut in 1967, and Mercury was acknowledging the fact and playing to its strengths.

A pair of engines were slotted into the new Cougar. The base powerplant is the Zetec, 16-valve inline, four-cylinder, iron-block, aluminum-head engine. With a 9.6:1 compression ratio, a 125 peak horsepower was achieved at 5,500 rpm, and torque was rated at 130 foot-pounds at 4,000 rpm. The 2.0-liter (122 ci) mill utilized Ford's EEC-V engine management system for optimal ignition timing and precise fuel mixture for maximum performance and good mileage. The all-aluminum Duratec 24-valve V-6, displacing 2.5-liters (155 ci) 24-valve Duratec powerplant was the hot optional engine. With a 9.7:1 compression ratio, this sand-cast alloy engine boasts 170 horsepower at a high 6,250 rpm and a 165 foot-pounds of torque arrives at 4,250 rpm.

This front-wheel-drive vehicle has two transmissions options—a Ford CDE dual-mode four-speed automatic with overdrive and a Ford MTX75 manual five-speed. The 106.4-inch wheelbase vehicle is fitted with independent suspension on both ends for sensational road handling. For maximum braking performance, the V-6 optioned cars come with four-wheel disc brakes. Interestingly, the new Cougar is only 5 inches shorter overall and 1.5 inches narrower than a 1967 Cougar. The New Edge design Cat moves from 0 to 60 in 7.7 seconds with the V-6/five-speed manual setup. In comparison, the 289-ci-engined Cougar tested in its January 1967 issue of *Motor Trend* took 7.0 seconds to scamper from 0 to 60. The quarter-mile is covered in 16.0 seconds in the 1999 Cougar at a speed of 87.3 miles per hour according to September 1998's *Motor Trend*. The *MT* crew took a 390-ci 1967 Cougar through the quarter-mile in 16.0 seconds at 89 miles per hour in their January 1967 issue. The more things change. . . the more they stay the same.

Next Page
The Cougar S concept car was introduced at the 1999 Los Angeles Auto Show. Cougar S designer Troy Trinh wanted to make sure the "S" was "more buff, a look to emphasize the power." The 3.0-liter V-6 provides scintillating acceleration to the all-wheel-drive car.

With the addition of Enkei SST-2 alloy wheels, 17x8 in the front, 18x8 in the rear, the aggressive stance of the Mercury Marauder concept Cat demonstrates that Mercury has not forgotten its performance past. *Ford Motor Company*

The Mercury Chief's Unique Vision: Jack Telnack

Jack Telnack swears that he was born with a blue oval on his rear; his record of employment would lend credence to that statement. Born at Henry Ford Hospital in Detroit, Michigan, this son of a Ford employee fell in love with the design of automobiles when he saw the 1941 Continental. Acknowledged as a masterful design to this day, Telnack modified his first car, a 1941 Mercury, to look like the high-end vehicle. He remembers, "I channeled the body to drop 6 inches down over the frame rails. I even lowered the cowl (firewall) 6 inches. That's what they did in the Continental. I drove that car to the Art Center in Pasadena when I was accepted there. By the time I got to California, the bottom of the steering wheel had worn a hole in the leg of my pants. But man, I looked cool!"

Jack Telnack, vice-president of design, Ford Motor Company, retired in 1998. He oversaw the implementation of both the Aero look and New Edge and helped forge Mercury's revitalized interest in high performance. *Ford Motor Company*

After graduation, it was back to Detroit, where he went to work for Arnott B. "Buzz" Grisinger in the Mercury studios. A succession of jobs in the Ford Motor Company design studios around the world eventually led to becoming vice-president of design, responsible for the sheet metal of Ford automobiles world-wide, including Lincoln-Mercury. He was the key individual that introduced the Aero look in 1986 with the Ford Taurus/Mercury Sable. When asked what he felt his best design of his career was, he responds, "I am the most proud of the Sable. It has traditional Mercury styling cues, from the vertical grille on the front bumper to the repeating vertical lines on the rear, like a 1967 Cougar." The Taurus/Sable helped Ford when it really needed the assistance. It opened the door for other manufacturers to incorporate aerodynamics into a production road car design.

Brand identity is often lost when vehicles share a common platform. Telnack fought in many meetings to preserve a separate Mercury identity. "Some of the top executives kept saying 'we want a Mercury flavor.' I tried to show them that Mercury had a clear design heritage that could be traced back to 1939. They just had to look!" In hindsight, he feels that maybe more could have been done. "If I could go back, I would have fought even harder for a design expression of Mercury's history. But with the constraints that we were working under, I think that we gave the Mercury line a different tone than Ford or Lincoln."

The Eliminator's Triumphant Return

The Cougar has a performance background. From 1974 to 1997, its performance was not stressed. After a performance version of the 1999 Cougar came out, the media inevitably asked about Mercury's high-performance destiny. The corporation line was that there were no plans for a stronger Cougar. But as Cougar designer Daryl Behmer pointed out, Mercury is going to be keeping the car fresh, styling as well as engineering-wise. "Since the twin-overhead cam V-6 is already fitted to the Contour platform, it's only natural for that engine to find its way into the Cougar."

In the same vein, Mercury displayed a concept car with a name from the past (Eliminator) at the annual Specialty Equipment Market Association (SEMA) show in Las Vegas in late 1998. With a Cougar lowered 1.5 inches, Ford's Special Vehicle Engineering group modified a 2.5-liter V-6 Cougar in a very performance-friendly fashion. The concept car was an exercise by designer Ken Grant as a way of showing owners what could be done with a Cougar using off-the-shelf parts. However, Mercury personnel didn't deny that public interest in the project could not hurt the chance of production.

The gang in Dearborn creation had a Vortech supercharger, K&N air cleaner, a specially fabricated Borla exhaust, Brembo disc brakes with drilled rotors filled huge 18-inch Enkei alloy wheels, 225/40R-18 Goodyear Eagle tires. The body was fitted with a ground-effect package from Inform Designs, and painted with a special PPG Chrom-O-Flair multi-hue finish. Chances for production seem good, especially when taken with Daryl Behmer's statements regarding the use of the bigger V-6. It's clear Mercury has not forgotten its performance past.

Old Name, New Face—the 1999 Marauder

Mercury had more surprises up its sleeve at the SEMA show in Las Vegas. It reintroduced another nameplate from the past—Marauder. Like the vehicles wearing this moniker in the 1960s, Mercury took a full-size vehicle, injected some power under the hood, and spruced up the body. It worked then and it works now. The engineers started with a 1999 Grand Marquis, a full-sized, comfortable Interstate cruiser. Under the hood, a Ford SVO-supercharged 4.6 SOHC V-8 was installed along with K&N air filters. A custom-fabricated 2.25-inch dual exhaust system managed the spent fuel charge. The revamped suspension featured Edelbrock Performer IAS units. Enkei SST-2 alloy wheels, 17x8 in the front, 18x8 in the rear, covered with Pirelli P-Zero tires provided a quantum leap in road-holding capability. Bucket seats replaced the front bench unit, and the exterior was coated with a monochromatic black paint. As this is written, Mercury has not yet spoken about any production plans. From a performance enthusiast's standpoint, this would be another way to show that Mercury intends to build vehicles that actually are fun to point down the road.

A Bold and Brash Cougar S

The future of Mercury performance is looking very bright. At the Los Angeles Auto Show in January 1999, Mercury took the wraps off yet another concept—the Cougar S. Designed by Troy Trinh in the Advanced Studio and built by MSX in Dearborn, Michigan, the Cougar S was another showcase intended to bring a younger clientele into Mercury showrooms. Mike Jennings, Cougar brand manager, pointed out that "Cougar S is very important for the exploding market of young car enthusiasts, who are increasingly looking to modify front- or all-wheel-drive imports with smaller engines."

The Cougar S contains Mercury's first all-wheel-drive (AWD), and it bristles with exciting technology. The drivetrain was lifted from the AWD Mondeo in Europe. The track was widened 3 inches front and rear, and it was lowered 1 inch. The 3.0-liter, 24-valve Duratec V-6 engine rated at 215 horsepower was pulled from the SVT Contour. It is hooked up to a five-speed MTX-75 transmission with a viscous-drive limited slip differential, and the final drive ratio is 4.06. Massive Brembo disc brakes with four-piston calipers up front and two-piston calipers in back provide magnificent stopping power and fill the 18-inch wheels.

The body modifications were designed by Trinh to be "sophisticated, tasteful, aggressive yet subtle." The larger grille opening was needed to allow sufficient cooling air for the higher horsepower engine, and the dual rear spoilers increase downforce. The Cougar S was designed from the inception to paper to finished product in 100 days. Trinh and supervisor Bob Barnes put enough hours in the studio to qualify it as their permanent residence, but deadlines required it.

The Cougar S hinted at the direction that Mercury is taking with performance. J. Mays, vice president of design, recently unveiled a five-year plan to develop a "visual language" that would differentiate each of Ford's brands. Mercury will define its products as "innovative, expressive, individualistic." How this will translate into exciting vehicles is something all enthusiasts are wondering. But the infusion of new, young talent at the company bodes well for performance. We seem to be living in another Golden Age. One thing the new leaders at Mercury realize is that speed and styling sells, and that Mercury has a performance past rich in both.

Index